TRANSPORTATION INFRASTRUCTURE - ROADS, HIGHWAYS, BRIDGES, AIRPORTS AND MASS TRANSIT

RED LIGHT CAMERA SYSTEMS

TRAFFIC SAFETY EVALUATION AND OPERATIONAL GUIDELINES

TRANSPORTATION INFRASTRUCTURE - ROADS, HIGHWAYS, BRIDGES, AIRPORTS AND MASS TRANSIT

Additional books in this series can be found on Nova's website under the Series tab.

Additional e-books in this series can be found on Nova's website under the e-book tab.

TRANSPORTATION INFRASTRUCTURE - ROADS, HIGHWAYS, BRIDGES, AIRPORTS AND MASS TRANSIT

RED LIGHT CAMERA SYSTEMS

TRAFFIC SAFETY EVALUATION AND OPERATIONAL GUIDELINES

CYNTHIA MUNOZ
EDITOR

nova publishers
New York

NOTICE TO THE READER

Library of Congress Cataloging-in-Publication Data

ISBN: 978-1-63483-448-3

Published by Nova Science Publishers, Inc. † New York

CONTENTS

PREFACE

Red light running is one of the major causes of crashes, deaths, and injuries at signalized intersections. The Federal Highway Administration (FHWA) and National Highway Traffic Safety Administration (NHTSA) support a comprehensive approach to intersection safety that incorporates engineering, education, and enforcement countermeasures to prevent red light running and improve intersection safety. Red light cameras can be a very effective countermeasure to prevent red light running. There are a number of studies that indicate reduction in crashes at signalized intersections due to red light cameras. The fundamental objective of the research in this book is to determine the effectiveness of RLC systems in reducing crashes. A description of all project efforts is described in this book.

In: Red Light Camera Systems
Editor: Cynthia Munoz

ISBN: 978-1-63483-448-3

Chapter 1

SAFETY EVALUATION OF RED-LIGHT CAMERAS*

Forrest M. Council, Bhagwant Persaud, Kimberly Eccles, Craig Lyon and Michael S. Griffith

I. EXECUTIVE SUMMARY

Abstract

The fundamental objective of this research was to determine the effectiveness of red-light- camera (RLC) systems in reducing crashes. The study involved an empirical Bayes (EB) before- after research using data from seven jurisdictions across the United States to estimate the crash and associated economic effects of RLC systems. The study included 132 treatment sites, and specially derived rear end and right-angle unit crash costs for various severity levels. Crash effects detected were consistent in direction with those found in many previous studies: decreased right-angle crashes and increased rear end ones. The economic analysis examined the extent to which the increase in rear end crashes negates the benefits for decreased right-angle crashes. There was indeed a modest aggregate crash cost benefit of RLC systems. A disaggregate analysis found that greatest economic benefits are

* This is an edited, reformatted and augmented version of a report, FHWA-HRT-05-048, issued by the U.S. Department of Transportation, Federal Highway Administration, April 2005.

associated with factors of the highest total entering average annual daily traffic (AADT), the largest ratios of right-angle to rear end crashes, and with the presence of protected left-turn phases. There were weak indications of a spillover effect that point to a need for a more definitive, perhaps prospective, study of this issue.

Introduction and Background

RLC systems are aimed at helping reduce a major safety problem at urban and rural intersections, a problem that is estimated to produce more than 100,000 crashes and approximately 1,000 deaths per year in the United States [1]. The size of the problem, the promise shown from the use of RLC systems in other countries, and the paucity of definitive studies in the United States established the need for this national study to determine the effectiveness of the RLC systems jurisdiction- wide in reducing crashes at monitored intersections. This study included collecting background information from literature and other sources, establishing study goals, interviewing and choosing potential study jurisdictions, and designing and carrying out the study of both crash and economic effects. A description of all project efforts is described in this report and, to a lesser extent, in two Transportation Research Board (TRB) papers [2, 3] that were also prepared.

A literature review found that estimates of the safety effect of red-light-running programs vary considerably. The bulk of the results appear to support a conclusion that red light cameras reduce right-angle crashes and could increase rear end crashes; however, most of the studies are tainted by methodological difficulties that would render useless any conclusions from them. One difficulty, failure to account for regression to the mean[1] (RTM), can exaggerate the positive effects, while another difficulty, ignoring possible spillover effects[2] to intersections without RLCs, will lead to an underestimation of RLC benefits, more so if sites with these effects are used as a comparison group.

While it is difficult to make definitive conclusions from studies with failed methodology validity, the results of the review did provide some level of

[1] "Regression to the mean" is the statistical tendency for locations chosen because of high crash histories to have lower crash frequencies in subsequent years even without treatment.

[2] Spillover effect is the expected effect of RLCs on intersections other than the ones actually treated because of jurisdiction-wide publicity and the general public's lack of knowledge of where RLCs are installed.

comfort for a decision to conduct a definitive, large-scale study of installations in the United States. It was important for the new study to capitalize on lessons learned from the strengths and weaknesses of previous evaluations, many of which were conducted in an era with less knowledge of potential pitfalls in evaluation studies and methods to avoid or correct them.

The lessons learned required that the number of treatment sites be sufficient to assure statistical significance of results, and that the possibility of spillover effects be considered in designating comparison sites, perhaps requiring a study design without a strong reliance on the use of comparison sites. Previous research experience also pointed to a need for the definition of the term, "red-light-running crashes," to be consistent, clear, and logical and for provision of a mechanism to aggregate the differential effects on crashes of various impact types and severities.

Methodological Basics

The general crash effects analysis methodology used is different from those used in past RLC studies. This study benefits from significant advances made in the methodology for observational before-after studies, described in a landmark book by Hauer [4] The book documented the EB procedure used in this study. The EB approach sought to overcome the limitations of previous evaluations of red-light cameras, especially by properly accounting for regression to the mean, and by overcoming the difficulties of using crash rates in normalizing for volume differences between the before and after periods.

The analysis of economic effects fundamentally involved the development of per-crash cost estimates for different crash types and police-reported crash severities. In essence, the application of these unit costs to the EB crash frequency effect estimates. The EB analysis was first conducted for each crash type and severity and site before applying the unit costs and aggregating the economic effect estimates across crash types and severity and then across jurisdictions. The estimates of economic effects for each site allowed for exploratory analysis and regression modeling of cross-jurisdiction aggregate economic costs to identify the intersection and RLC program characteristics associated with the greatest economic benefits of RLC systems.

Details of the development of the unit crash-cost estimates can be found in a recent paper and in an internal report available from FHWA [5, 6]. Unit costs were developed for angle, rear end, and "other" crashes at urban and rural signalized intersections. The crash cost to be used had to be keyed to police crash severity based on the KABCO[3] scale. By merging previously developed costs per victim keyed on the AIS injury severity scale into U.S. traffic crash data files that scored injuries in both the Abbreviated Injury Scale (AIS) and KABCO scales, estimates for both economic (human capital) costs and comprehensive costs per crash were produced. In addition, the analysis produced an estimate of the standard deviation for each average cost. All estimates were stated in Year 2001 dollar costs.

Data Collection

The choice of jurisdictions to include in the study was based on an analysis of sample size needs and the data available in potential jurisdictions. It was vital to ensure that enough data were included to detect that the expected change in safety has appropriate statistical significance. To this end, extensive interviews were conducted for several potential jurisdictions known to have significant RLC programs and a sample size analysis was done. The final selection of seven jurisdictions was made after an assessment of each jurisdiction's ability to provide the required data. The jurisdictions chosen were El Cajon, San Diego, and San Francisco, CA; Howard County, Montgomery County, and Baltimore, MD; and Charlotte, NC.

Data were required not only for RLC-equipped intersections but also for a reference group of signalized intersections not equipped with RLCs but similar to the RLC locations. These sites were to be used in the calibration of safety performance functions (SPFs) used in the EB analysis and to investigate possible spillover effects. To account for time trends between the period before the first RLC installation and the period after that, crash and traffic volume data were collected to calibrate SPFs from a comparison group of approximately 50 unsignalized intersections in each jurisdiction.

Following the site/jurisdiction selection, the project team collected and coded the required data. Before the actual data analyses, preliminary efforts involving file merging and data quality checks were conducted. This effort included the crash data linkage to intersections and the defining of crashes expected to be affected by RLC implementation. Basic red-light-running crashes at the intersection proper were defined as "right-angle," "broadside,"

---

[3] The KABCO severity scale is used by the investigating police officer on the scene to classify injury severity for occupants with five categories: K, killed; A, disabling injury; B, evident injury; C, possible injury; O, no apparent injury. (7) These definitions may vary slightly for different police agencies.

or "right- or left- turning-crashes" involving two vehicles, with the vehicles entering the intersection from perpendicular approaches. Crashes involving a left-turning vehicle and a through vehicle from opposite approaches were also included. "Rear end crashes" were defined as a rear end crash type occurring on any approach within 45.72 m (150 ft) of the intersection. In addition, "injury crashes" were defined as including fatal and definite injuries, excluding those classified as "possible injury."

Results

Because the intent of the research was to conduct a multijurisdictional study representing different locations across the United States, the aggregate effects over all RLC sites in all jurisdictions was of primary interest. A significant decrease in right-angle crashes was found, but there is also a significant increase in rear end crashes. Note that "injury" crashes are defined by severity as K, A, or B crashes; but the frequencies shown do not contain a category for "possible injury" crashes captured by KABCO-level C; thus, these crashes could better be labeled "definite injury" crashes.

These effects, the direction if not the order of magnitude, were remarkably consistent across the jurisdictions. The analysis indicated a modest spillover effect on right-angle crashes; however, that this was not mirrored by the increase in rear end crashes seen in the treatment group, which detracts somewhat from the credibility of this result as evidence of a general deterrence effect.

For the analysis of economic effects, it was recognized that there were low sample sizes of fatal and serious (A-level) crashes in the after period for some intersections. In addition, the initially developed cost estimates for B- and C-level rear end crashes indicated some anomalies in the order (e.g., C-level costs were higher, very likely because on-scene police estimates of "minor injury" often ultimately include expensive whiplash injuries), the B- and C-level costs were combined by Pacific Institute for Research and Evaluation (PIRE) into one cost. Considering these issues and the need to use the same cost categories across all intersections in all seven jurisdictions, two crash cost levels were ultimately used in all analyses: Injury (K+A+B+C) and Noninjury (O).

The study summarizes results for the economic effects including and excluding property-damage only (PDO) crashes. The latter estimates are included in recognition of the fact that several jurisdictions considerably under-report PDO collisions. Those estimates (with PDOs excluded) show a positive aggregate economic benefit of more than $18.5 million over approximately 370 site years, which translates into a crash reduction benefit of approximately $50,000 per site year. With PDOs included the benefit is approximately $39,000 per site year. The implication from this result is that the lesser severities and generally lower unit costs for rear end injury crashes together ensure that the increase in rear end crash frequency does not negate the decrease in the right-angle crashes targeted by red-light-camera systems.

Further analysis indicated that right-angle crashes appear slightly more severe in the after period in two jurisdictions, but not in the other five. Because such an effect would cause a slight overestimation in economic benefits, an attempt was made to estimate the possible size of the benefit reduction. If such a shift were real, and if its effects could be assumed to be correctly estimated from individual KABCO unit costs already deemed to be inappropriate for such purposes, the overall cost savings reported could be decreased by approximately $4 million; however, there would still be positive economic benefits, even if it is assumed that the unit cost shifts were real and correctly estimated.

Examination of the aggregate economic effect per after-period year for each site indicated substantial variation, much of which could be attributable to randomness. It was reasonable to suspect that some of the differences may be due to factors that impact RLC effectiveness; therefore, a disaggregate analysis, which involved exploratory univariate analysis and multivariate modeling was undertaken to try to identify factors associated with the greatest and least economic benefits. The outcome measure in these models was the aggregate economic effect per after period site year.

The disaggregate analysis found that greatest economic benefits are associated with the highest total entering AADTs, the largest ratios of right-angle to rear end crashes, higher proportions of entering AADT on the major road, shorter cycle lengths and intergreen periods, and with the presence of protected left-turn phases. The presence of warning signs and high publicity levels also appear to be associated with greater benefits. These results do not provide numerical guidance for trading off the effects of various factors. The intent of identifying these factors is that in practice RLC implementers would identify program factors such as warning signs that increase program effectiveness and give the highest priority for RLC implementation to the sites with most or all of the positive binary factors present (e.g., left-turn protection) and with the highest levels of the favorable continuous variables (e.g. higher ratios of right-angle to rear end crashes).

Conclusion

This statistically defendable study found crash effects that were consistent in direction with those found in many previous studies, although the positive effects were somewhat lower that those reported in many sources. The conflicting direction effects for rear end and right-angle crashes justified the conduct of the economic effects analysis to assess the extent to which the increase in rear end crashes negates the benefits for right-angle crashes. This analysis, which was based on an aggregation of rear end and right-angle crash costs for various severity levels, showed that RLC systems do indeed provide a modest aggregate crash-cost benefit.

The opposing effects for the two crash types also implied that RLC systems would be most beneficial at intersections where there are relatively few rear end crashes and many right-angle ones. This was verified in a disaggregate analysis of the economic effect to try to isolate the factors that would favor (or discourage) the installation of RLC systems. That analysis revealed that RLC systems should be considered for intersections with a high ratio of right-angle crashes to rear end crashes, higher proportion of entering AADT on the major road, shorter cycle lengths and intergreen periods, one or more left turn protected phases, and higher entering AADTs. It also revealed the presence of warning signs at both RLC intersections and city limits and the application of high publicity levels will enhance the benefits of RLC systems.

The indications of a spillover effect point to a need for a more definitive study of this issue. That more confidence could not be placed in this aspect of the analysis reflects that this is an observational retrospective study in which RLC installations took place over many years and where other programs and treatments may have affected crash frequencies at the spillover study sites. A prospective study with an explicit purpose of addressing this issue seems to be required.

In closing, this economic analysis represents the first attempt in the known literature to combine the positive effects of right-angle crash reductions with the negative effects of rear end crash increases and identify factors that might further enhance the effects of RLC systems. Larger crash sample sizes would have added even more information. The following primary conclusions are based on these current analyses:

1. Even though the positive effects on angle crashes of RLC systems is partially offset by negative effects related to increases in rear end crashes, there is still a modest to moderate economic benefit of between $39,000 and

$50,000 per treated site year, depending on consideration of only injury crashes or including PDO crashes, and whether the statistically non-significant shift to slightly more severe angle crashes remaining after treatment is, in fact, real.

2. Even if modest, this economic benefit is important. In many instances today, the RLC systems pay for themselves through red-light-running fines generated. However, in many jurisdictions, this differs from most safety treatments where there are installation, maintenance, and other costs that must be weighed against the treatment benefits.

3. The modest benefit per site is an average over all sites. As the analysis of factors showed, this benefit can be increased through careful selection of the sites to be treated (e.g., sites with a high ratio of right-angle to rear end crashes as compared to other potential treatment sites) and program design (e.g., high publicity, signing at both intersections and jurisdiction limits).

Statistical information related to this Executive Summary can be found in tables 1 through 4 in this final report.

II. Introduction

A red-light-camera (RLC) system automatically detects when a vehicle has entered an intersection during the red phase for an approach and takes a photograph of the red-light-running violation. Jurisdictional staff members review photographs to determine if a ticket should be sent to the driver.

This treatment (i.e., crash countermeasure) is aimed at helping reduce a major safety problem at urban and rural intersections. Red-light-running is estimated to produce over 95,000 crashes and approximately 1,000 deaths per year [1]. After being used extensively overseas for over a decade, the use of RLC systems has risen dramatically in the U.S. in recent years. Given the size of the problem, and the promise shown from the use of RLC systems in other countries (and by some studies in the U.S.), there is a clear need to determine the effectiveness of the RLC systems in reducing crashes at monitored intersections as well as jurisdiction-wide. There is also a need to determine if such programs can be made more cost-effective through changes in such variables as signage, signal phasing, and public information programs. Studies conducted in over a dozen U.S. cities and several foreign countries indicate that RLC programs are effective in reducing the number of red-light-running *violations*. However, there is much less evaluation-based knowledge on the effect of RLC programs on *crashes*, especially the kinds of crashes typically caused by red-light running.

While less controversial than speed-enforcement camera programs, RLC programs are not entirely without controversy. Even though an apparent safety benefit has been indicated in some studies, such programs have advocates and detractors at both local community levels and as high as the U.S. Congress. The debate is fuelled by the reality that no well-conducted, scientifically-sound, multijurisdiction evaluation of RLC program effectiveness in reducing crashes has been undertaken.

To meet this need, the Federal Highway Administration (FHWA) ITS (intelligent transportation system) Joint Program Office (JPO) requested submission of an RLC evaluation proposal under the ITS (intelligent transportation system) Program Assessment Support contract. The resulting contract supported the work described in this report.

III. Project Overview

As shown in figure 1, the work was conducted in two phases. Phase I involved the development of a detailed experimental design. This was reviewed and subsequently approved by FHWA with minor modifications. In Phase II, the design was then implemented. This report describes both of these phases in detail.

Phase I—Evaluation Design

As shown in figure 1, the Phase I effort involved multiple steps to establish goals for the evaluation, collection of background information that would help shape the evaluation plan, choice of the jurisdictions to be involved in the evaluation, and definition of the data collection and statistical techniques to be used. Following are the tasks in the Phase I experimental design:

- *Conduct literature review*—The study team specified, obtained, and produced detailed critical reviews of key U.S. and international studies. The intent of this review was not only to summarize what is known concerning program effectiveness, but also to identify the critical experimental design factors that can overcome problems found in past studies.

- *Determine study questions*—The study team worked with the oversight panel established by FHWA to define and prioritize the study questions to be answered by the experimental design. The primary effort was in a workshop held early in the contract period, and the preliminary results of the literature review were used in the decisions.

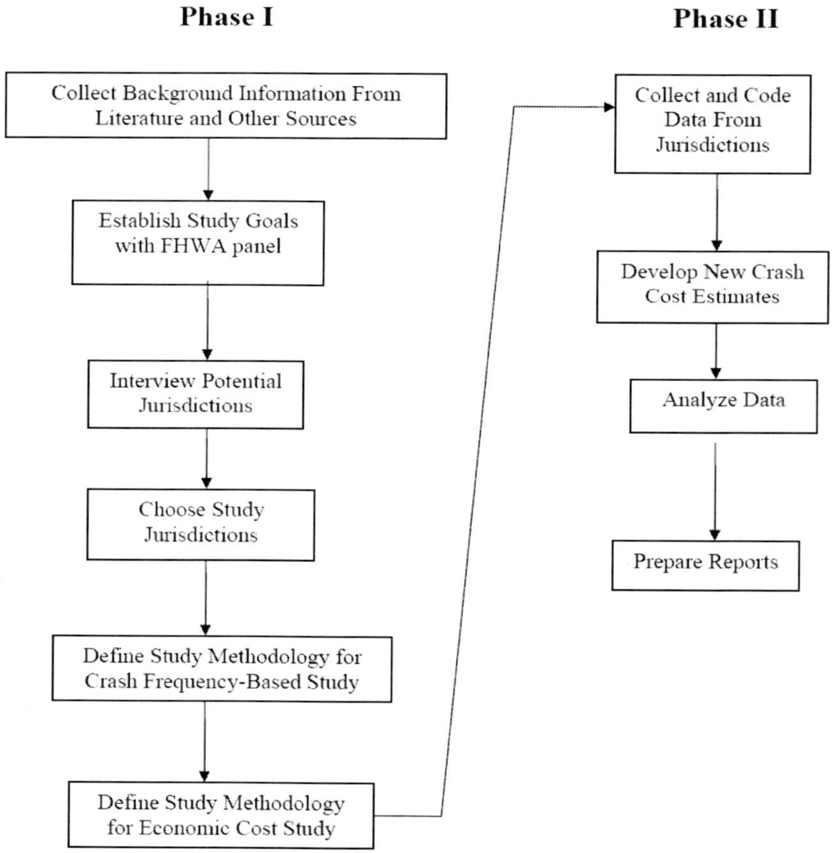

Figure 1. Project workflow.

- *Determine RLC-related data availability*—The project team conducted telephone interviews with 15 local agencies that have significant RLC programs in place. The most important questions to be answered concerned not only the details of the RLC program (e.g., number of intersections, violation definition, level of fine), but also the quality and availability of data related to before-and after-program crashes, signal data for treated and untreated intersections, traffic flows, and public information efforts.
- *Specify a multijurisdictional experimental design to estimate RLC safety effects*—Based on the outputs from the preceding tasks, the study team defined the multijurisdictional experimental plan, including data needs, the most appropriate local agencies to work with, data analysis techniques, and the specification of which of the desirable study questions might be answerable.
- *Specify an experimental plan for examining the effects of RLC on economic costs of related crashes*—Because RLCs are likely to decrease more severe angle crashes and increase less severe rear end crashes, the study team also defined a study method which can combine both crash frequency and severity in terms of economic costs.

As the task list shows, there were actually two experimental plans or methodologies developed in this Phase I effort. The first is a national design in which data from multiple jurisdictions across the United States were used to both define the crash-related effectiveness of RLC programs and to explore factors that could make such programs more beneficial in terms of crash reduction (e.g., signalization variables such as clearance intervals and cycle lengths, and public information programs and signage related to RLC programs). The primary outcome variable of interest here is RLC-related change in the frequencies of different types of crashes. The second experimental plan involved the development of a database and methodology to examine, in economic terms, costs and benefits of RLCs, thus simultaneously analyzing changes in crash severity and crash frequency. The following sections describe the development of this economic analysis methodology initiated in Phase I and revised and completed in Phase II; The full development process is described in the Phase II section.

Phase II—Evaluation Implementation

After FHWA agreed with the Phase I evaluation design, funding was provided for the implementation of the evaluation efforts. The project team then performed the following steps:

- *Collect and code data from jurisdictions*—Project staff established contacts with and visited the seven jurisdictions chosen for participation. Staff collected data related to intersection geometry, traffic flow, signalization parameters, and other descriptive data for treatment, signalized reference, and unsignalized reference intersections in each jurisdiction. The data, almost always in noncomputerized form, were then collected for data extraction, coding, and analysis file preparation. Because crash data for a multiyear before-period were not available in six of the seven jurisdictions, crash data were located and extracted from State data files.
- *Develop new crash cost estimates*—The final economic analysis methodology required the development of human capital and comprehensive cost estimates for each crash severity level in each pertinent crash type (e.g., the cost of a "moderate injury" in a rear end collision at a signalized intersection). Because FHWA cost data had not been updated since 1994, and because even those older data were not specific to crash and location types, a subcontractor with extensive experience in highway safety economics joined the team to develop these estimates.
- *Analyze data*—Following the national-study design and the crash-cost methodologies developed in Phase I, the project team analyzed both RLC-related crash frequency and crash cost in each of the seven jurisdictions. The effectiveness estimates from all jurisdictions were then combined to develop a national estimate of the effectiveness of RLC system.
- *Prepare reports*—Based on the analysis results, project staff then prepared this FHWA final report, an FHWA technical summary, and two Transportation Research Board (TRB) journal articles (one on the crash effects analysis, the other on the analysis of economic effects and the identification of factors associated with the greatest economic benefits) [2,3].

The narrative following the literature review will describe each of the steps in both Phase I and Phase II in more detail.

IV. Literature Review of Critical Studies

Unlike many literature reviews for safety research efforts, the goal of this task was not to review a large number of studies to summarize findings on RLC program effectiveness. Because the overall Phase I goal was to produce a scientifically sound experimental plan that could overcome as many threats to validity as possible, the literature review was aimed at a shorter list of international studies judged by the study team and the oversight panel to be critical studies. To this end, all possible studies of relevance were first identified on the basis of Internet searches such as Transportation Research Information Services (TRIS), and information from parallel and recent reviews and meta-analyses conducted for FHWA, the National Cooperative Highway Research Program (NCHRP), and the Insurance Institute for Highway Safety. The final choice of critical studies included studies from both the United States and other countries with a longer history of RLC program implementation, studies that appeared to be best in terms of scientific rigor, and studies often cited by other researchers or in political discussions of RLC effectiveness. The team scanned a number of study sources and reports and ultimately defined a listing of 17 critical studies.

A study team member then reviewed each of these studies in detail. The goal was to not only extract information on measured RLC program effectiveness, but also identify problems or issues that we would attempt to overcome in this new evaluation design. To accomplish this, listings of study strengths and weaknesses were developed for each study reviewed.

In the sections that follow, a general summary of the literature findings is presented first, followed by an itemization of the lessons learned from this exercise.

Summary of Findings

The studies reviewed varied widely, including the following areas:

- Accident types (all, right-angle, those caused by red-light running).
- Accident severities (all, injury plus fatal, weighted).

- Area of study (treated intersections, treated approaches, jurisdiction-wide).
- Use and designation of comparison sites.
- Treatment type (cameras only, cameras plus warning signs, red-light-running and speed cameras).
- Sample sizes, ranging from 3 to 78 camera-equipped intersections.
- Countries (several from Australia and the United Kingdom, but few from the United States).
- Study methodology (simple before-and-after, before-and-after with comparison group, chi-squared tests, statistical modeling).

It is not surprising that estimates of the safety effect of cameras vary considerably. A summary of the more relevant study findings is provided in table 1, including a synopsis of the main difficulties.

From table 1, one could conclude that the bulk of the results support a conclusion that red-light cameras reduce right angle crashes and could increase rear end crashes; however, as the last column shows, most studies are tainted by methodological difficulties that raise questions about any conclusions from them. One difficulty, failure to account for regression to the mean, can exaggerate the positive effects, while another, ignoring possible spillover effects at intersections without RLC, will lead to an underestimation of RLC benefits, even more so if sites with these effects are used as a comparison group. ("Spillover effect" is the expected effect of RLCs on intersections other than the ones actually treated, resulting from jurisdiction-wide publicity and the general public's lack of knowledge of where RLCs are installed.) Almost all studies had one or the other of these flaws and many had both, in addition to other flaws.

A similar assessment of the literature was made independently in a recent meta-analysis, in which the review for the Insurance Institute for Highway Safety included most of the same studies cited in table 1 and some others [23]. That work found, expectedly, that largest safety benefits were reported by studies that did not control for regression to the mean and that small effects tend to be found where the possibility of spillover was ignored. The one study that measured both spillover and specific effects, while ensuring that regression to the mean was not a factor, was an evaluation of the Oxnard, California program by the Insurance Institute for Highway Safety [16]. That study found a significant reduction in injury crashes overall but did not separate the specific effects at treatment sites from citywide effects. (It is understood that a follow up study is doing this.)

Table 1. Summary of findings from past studies

Reference	City	Camera sites	Comparison/ reference group	Crash type studied and estimated effects (negative indicates reduction)		Comment
Hillier, et al. (1993)(8)	Sydney, Australia	Installed at 16 intersections	16 signalized intersections	Right-angle and left-turn opposed	-50%	RTM* possible; spillover may have affected comparison sites; results confounded by adjustment to signal timing in middle of study period
				Rear end	+25% to 60%	
South, et al. (1988)(9)	Melbourne, Australia	Installed at 46 intersections	50 signalized intersections	No significant results. Looked at right angle, right-angle (turn), right against thru, rear end, rear end (turn), other, all crashes, number of casualties, no significant results		RTM* possible, no accounting for changes in traffic volumes; comparison sites possibly affected by spillover and other treatments
Andreassen (1995)(10)	Victoria, Australia			No significant results		Lack of an effect could be that the sites studied tended to have few red-light-running related accidents; comparison sites may have been affected by spillover
Kent, et al. (1995)(11)	Melbourne, Australia	3 intersection approaches at different intersections	Noncamera approaches	No significant relationship between the frequency of crashes at RLC and non-RLC sites and differences in red-light- running behavior		Cross-sectional design is problematic; likely spillover effects to the noncamera approaches at the same intersections

Table 1. (Continued)

Reference	City	Camera sites	Comparison/ reference group	Crash type studied and estimated effects (negative indicates reduction)		Comment
Mann, et al. (1994)(12)	Adelaide, Australia	Installed at 13 intersections	14 signalized intersections	Reductions at the camera sites were not statistically different from the reductions at the comparison sites		RTM*and spillover to comparison sites are issues not addressed
London Accident Analysis Unit (1997)(13)	London, U.K.	RLC at 12 intersections and 21 speed cameras	Citywide effects examined	No significant results		The results are confounded because two programs are evaluated
Hooke, et al. (1996)(14)	Various cities in England and Wales	Installed at 78 intersections		All injury	-18%	A simple before-and-after comparison not controlling for effects of other factors, RTM* and traffic volume changes; therefore there is limited confidence in the results.
Ng, et al. (1997)(15)	Singapore	Installed at 42 intersections	42 signalized intersections	All	-7%	RTM*and spillover effects at comparison sites are issues
				Right angle	-8%	
Retting and Kyrychenko (2001)(16)	Oxnard, CA	Installed at 11 intersections	Unsignalized intersections in Oxnard and signalized intersections in 3 similarly sized cities	All	-7%	Looked at citywide effects, not just at RLC sites 29 months of before-and- after data used
				All injury	-29%	
				Right angle	-32%	
				Right-angle injury	-69%	
				Rear end	+3%(nonsignificant)	

Reference	City	Camera sites	Comparison/ reference group	Crash type studied and estimated effects (negative indicates reduction)		Comment
SafeLight, Charlotte(17)	Charlotte, NC	Installed at 17 intersections	no comparison group	Angle—all approaches	-37%	Probable RTM in site selection
				Angle—camera approaches	-60%	
				All—camera approaches	-19%	
				Rear end—camera approaches	+4%	
				All	< -1%	
Maryland House of Delegates (2001)(18)	Howard County, MD	Installed at 25 intersections		Rear end	-32%	Probable RTM in site selection
				Right angle	-42%	
				Other	-22%	
Fleck and Smith (1998)(19)	San Francisco, CA	Installed at 6 intersections	Citywide effects examined	Citywide injury collisions caused by red-light violators; unclear how these were defined	- 9%	Question on definition of RLC crashes; did not examine specific effects at treated sites

Table 1. (Continued)

Reference	City	Camera sites	Comparison/ reference group	Crash type studied and estimated effects (negative indicates reduction)		Comment
Vinzant and Tatro (1999)[20]	Mesa, AZ	6 intersections with RLC only, 6 intersections with RLC plus photo speed enforcement	6 signalized intersections	Total crash rates—crashes per million entering vehicles at each intersection		It is unclear if the assignment of treatment/no treatment to the four quadrants was random
				Combined-treatment quadrant	- 15.9%	
				Photo-radar quadrant	- 7.5%	
				RLC quadrant	- 9.7%	
				Control quadrant	- 10.7%	
Fox (1996)[21]	Glasgow, Scotland	Installed at 8 intersections and 3 "pelican" crossings	Area wide effects on injury crashes examined	Crossing carelessly	- 54.0%	RTM effects likely because the decreases in non-RLR crashes are greater than the RLR decreases at times, it is difficult to say what citywide effect the cameras have.
				Unsafe right turn	- 29.0%	
				Failure to keep distance	+ 8.0%	
				Other	- 29.0%	
				All per month	- 32.0%	
Winn (1995)[22]	Glasgow, Scotland	6 locations on 1 approach	Various	Injury crashes related to RLR violations	- 62.0%	Probable RTM effects

* RTM = Regression to the mean, also called "bias by selection."

While it is difficult to make definitive conclusions from studies that generally fail the tests on the validity of the methodology, the results did provide some level of comfort for a decision to conduct a definitive large-scale study of U.S. installations. It was important, however, that the planned study capitalize on lessons learned from the strengths and weaknesses of the previous evaluations, many of which were conducted in an era when knowledge of potential pitfalls in evaluation studies and methods of avoiding or correcting them was not widespread. These lessons are reviewed next.

Lessons Learned and Issues Raised by Literature Search

From the literature review, a number of lessons were learned that were useful in designing a definitive U.S. study. Following is an itemization:

- *Number of treatment sites:* This was limited in many studies, making for low significance of many results. A definitive study will have to pay careful attention to sample sizes.
- *RLC "spillover effects" in same city:* Crashes could be affected at control/comparison sites, making it necessary to have such sites in other similar cities. This will also make it difficult to determine the effect at treated location versus all other locations in the same city, perhaps requiring a study design without a strong reliance on the use of comparison sites.
- *Differences in accident investigation and reporting practice between jurisdictions:* This will make intracity comparisons or amalgamating data difficult and may require a separate analysis of the more severe crashes that are less likely to be affected by these differences if amalgamation is necessary to achieve large enough sample sizes.
- *Defining "red-light-running crashes":* The lack of precise definition in past studies, the lack of clarity between angle and turning crashes on police forms, and the lack of information on "legal" right-turn-on-red crashes could cloud the definition of the outcome variable.
- *RLC effects on rear end crashes:* There is clearly a need to consider not only this crash type in the analysis, but also to account for the tradeoff in crash severity between right angle and rear end types. To do so requires the use of the economic cost of crashes as an outcome variable.

- *Exposure changes between before-and-after periods:* Exposure is the major determinant of intersection crashes. Therefore, it is important to account for any changes between the before and after period, particularly if these changes are triggered by the measure. All studies reviewed have failed to do this accounting, conveniently assuming that RLCs will not change exposure. It is also important to use proper methods for accounting for exposure changes because the conventional method of normalizing with crash rates (per unit of traffic volume) is a dubious one, given the non-linear relationship between crashes and exposure found in many cases.
- *Regression to the mean effects*: In almost all studies reviewed, RLCs were installed at intersections with a high incidence of crashes, particularly those likely to be affected by RLC; this can lead to significant regression to the mean, particularly in the positive effects on RLC-targeted crashes, which must be accounted for.
- *Yellow interval improvements (and other intersection improvements) made at time of installation of RLCs:* This makes it difficult to determine what caused the measured effect. It especially important to separate the effects of these other measures from that of RLC because some studies have shown that these other treatments can be just as effective as RLC. Thus the study should not resolve the difficulty of the confounding effects of other measures by avoiding affected sites, but should intentionally seek to include some of these sites if possible.
- *Disaggregate effects by signalization variables:* There is little knowledge on the effect of variables such as cycle length and yellow and all-red interval combinations. Such knowledge will be useful in planning RLC programs or in explaining differential effects across sites and jurisdictions. The ability to seek this knowledge will be affected by the size of the sample and the variation in these factors.
- *Effect of signage:* Warning the driver at a specific location may or may not change the effect, and may or may not limit the possible treatment effect on other intersections. Signing and enforcing one approach has been reported to have a benefit for all other approaches in some studies, but some studies have found otherwise. The issue will require further investigation, which will require that appropriate data be collected.
- *Public education level:* It is desirable that this be specified and measured and its effects evaluated, in the light of other enforcement

research that has shown the importance of combining enforcement with public information (PI) programs.

- *Type of ticketing:* This affects true enforcement level and driver perception of "cost" of a ticket. The level of the fine and whether there are driver points for a violation can affect outcome, and the requirement to ticket the driver (rather than owner) can mean that only low percentage of offenders (i.e., 25 percent in San Francisco) are ticketed. This issue requires resolution, which will require that appropriate data be collected.
- *Definition of Red-Light violation:* This could affect ticketing and public perception. Short or long "grace period" after signal turns red could have different effects. It would be desirable to isolate these differential effects, if possible.
- *Camera rotation:* Determining if one can have a greater effect with the same number of cameras rotated to more sites is important. However, the optimum amount of rotation is undefined, so it might be desirable to develop this knowledge.
- *Relationship between changes in violations and changes in crashes:* This will depend on many factors including the grace period chosen, driver versus owner, etc. To date, there is no knowledge on such a relationship. Establishing a link would be useful in that it would considerably simplify the task of evaluating RLC installations. This, however, will likely require a prospective study.

These "lessons learned" were then incorporated into the experimental designs for both the crash- frequency-based study and the economic analysis study covered in later sections of this report.

V. Determination of Study Questions to Be Answered

The core study question to be answered is, "*What effect does RLC programs have on intersection safety, as measured by changes in crashes*?" The evaluation design presented later has this question as its primary focus. However, even this question requires some further clarification and expansion, in that RLC installations at selected intersections in a jurisdiction are often part of a larger "Reduce Red-Light-Running" campaign that is jurisdiction-wide.

The overall program will include a public information component (whether planned or just as media coverage of this new enforcement technique) that can clearly have an effect on driving behavior at other intersections in the jurisdiction. Thus, the core question is immediately expanded to, "*What are both the local effect of RLCs at treated intersections, and the 'spillover' effect at nearby intersections or jurisdiction-wide?*" In addition, there are other program components or factors that might make such programs more beneficial in terms of crash reduction. Examples could include yellow interval or phasing changes done with camera installation, public information programs and signage related to RLC programs, and the issue of whether to ticket vehicle owners or vehicle drivers.

Because not every conceivable question can be answered in one evaluation, particularly a retrospective evaluation in which programs and data have been determined by the local agency rather than the evaluator, there was a need to clearly establish a list of key study questions to focus on (if the data would allow it). The FHWA established an internal project oversight panel to make these decisions. Forrest M. Council of BMI-SG and Bhagwant Persaud of Ryerson University, the project co-principal investigators, met with this panel in early January, 2002. Panel members attending that meeting included the following.

Michael Griffith, *Chair*, FHWA Office of Safety Research and Development
Pam Crenshaw, FHWA Office of Operations, Travel Management
Pat Hasson, FHWA Resource Center
Hari Kalla, FHWA Office of Safety, Safety Design
John McFadden, FHWA Resource Center
Joe Peters, ITS Joint Program Office
Amy Polk, NASA Jet Propulsion Laboratory
Greg Hatcher, Mitretek Systems, Inc.
Rob Maccubbin, Mitretek Systems, Inc.

The panel and project team discussed items such as issues related to the literature review (e.g., defining "critical studies," proposed study list, current progress, example detailed reviews), a listing of "lessons learned/issues raised" from the literature reviewed to date, preliminary thoughts on the experimental plan, etc. A major part of the discussion centered on the listing of study questions to be answered. At that initial meeting, the Chair and panel defined a draft listing of questions. This was revised slightly into the following final listing based on findings concerning available data and the range or spread of the data among the jurisdictions.

First-Level Priority

- What effect do RLCs have on intersection safety (i.e., intersection crashes) at monitored intersections versus intersection safety throughout the jurisdiction?
- What is the relationship of signal timing (i.e., length of the yellow interval, length of the all-red interval, and various combinations of the yellow interval and all-red interval) with safety at intersections with RLCs? Later discussion indicated that the key factors of interest are yellow interval, all-red interval, cycle length, and signal coordination. The basic issue related to yellow time is the nature of the yellow phase length—e.g., a standard length, length based on Institute of Transportation Engineers (ITE) recommendations related to approach speeds and other factors, or some variation of these. The basic question for all-red phases is whether or not there is one (i.e., presence or absence of all-red phase). Cycle length is needed both to provide some measure of the number of red phases (and thus the number *of opportunities* for red-light-running) in a given time period, but also because longer red phases might "induce" more red-light- running. With respect to signal coordination, the issue is whether the treated signal approach is part of a set of coordinated signals that lead to queuing of vehicles (but not any additional details of the level of coordination).
- Are there certain improvements (e.g., signal timing, signage, geometric changes, etc.) done in conjunction with RLC installation that make the automated enforcement program more or less effective? Later discussion of the signage issue indicated that the key question has to do with presence or absence of "warning" signage, whether the sign is located at the intersection or away from the intersection (e.g., at the edge of town or at the beginning of a corridor), and whether informational signs providing data to the public on the number of red-light-running violations that have been issued are installed (because such signs have been shown to increase the effect of seatbelt enforcement programs and perhaps RLC programs in some cities).

- If other improvements are made during the installation of RLCs, what portion of the change in intersection crashes is due to these improvements and what portion is due to the RLC?
- What effect does a "good" public information program have on safety at intersections with RLCs? Given the fact that public information has been shown to be an important part of other effective enforcement programs (e.g., belt-use programs and driving under the influence (DUI) roadblocks), there is a need for some measure of "public information program level" for the cities. As noted later, the city interview form did contain three such levels.

Second-Level Priority

- What is the effectiveness of "fine only" (i.e., owner liability) program versus "fine and points" (i.e., driver liability) program?
- What is the relationship between the reduction in violations and the reduction in crashes? *(Note that this issue is not covered in the proposed experimental design because it would require a somewhat different study approach; crash effects is the primary focus.)*
- What is the relationship of the grace period (i.e., ticketing threshold—how far into the red phase before a ticket is issued) and location of the camera activation loops with safety at intersections with RLCs? (*Note that this was later removed as a question of interest because the "lag time" is a function of how the camera system is installed along with the legal definition of a violation. Because the driver is not aware of the extent of the "grace period," it is no longer felt to be of significant interest.*)

VI. Determination of RLC-Related Data Availability

Part of the difficulty in defining specific study questions early in the project was the lack of information on data availability and range for the different issues. While the project team (and the FHWA panel) had some insight into what data might be available on a State-controlled roadway system, there was less knowledge about what would be available in city or county data systems. For this reason, a decision was made to interview a small sample of key cities across the U.S. to address these questions.

The "sampling frame" for these interviews was based on a November 2002, list provided by the ITS Joint Program Office. As part of their periodic survey of ITS deployment in 78 of largest metropolitan areas in the U.S., a set of questions asked about current and projected numbers of intersections where RLC systems were installed. An additional listing of jurisdictions with RLC programs was extracted from a Web page listing developed by the Insurance Institute for Highway Safety. Additional cities were added based on panel knowledge and on a separate study where RLC programs were being surveyed.

The initial idea was to conduct a survey of a random sample of jurisdictions in this combined listing to determine, in general, what data might be available. (Note that only limited funds had been set aside for this task at the inception of the project.) However, after the detailed interview form was developed and tested, it became apparent that the interview would require 30–50 minutes of time with pertinent city/county staff, that it was often difficult to identify and reach the appropriate staff (even with the ITS contacts on the JPO listing), and that staff in more than one city/county agency would usually have to be contacted due to the cross-agency nature of this program (i.e., both traffic and police staff). For these reasons, a decision was made to interview cities with 10 or more RLCs already in operation. Emphasis was also placed on cities that had a significant number of such treated locations in place in 1999 or earlier—sites that would be useful in a retrospective study requiring at least a 2-year "after" period. While "nonrandom," this provided information on cities that were potential future participants in the actual evaluation. Within the time frame and budget, project staff interviewed 15 jurisdictions.

- Chandler, AZ.
- Prince George's County, MD.
- Fairfax County, VA.
- El Cajon City, CA
- Montgomery County, MD.
- New York City, NY.
- City of San Diego, CA.
- San Francisco, CA.
- Boulder, CO.
- Sacramento City, CA
- Charlotte, NC.
- Arlington County, VA.

- Howard County, VA.
- Baltimore, MD.
- Greensboro, NC.

While most of these interviews were completed by phone (as were follow up questions), two jurisdictions chose to complete the interview in writing. This led to some missing data for a limited number of questions. However, in general, the interview process did produce detailed data on both the nature of the RLC program and the availability of various forms of data in each jurisdiction. (The interview form and a summary of the results are available from the authors on request.) The results of the interviews will be provided at the end of the following section.

VII. Methodology for National, Multijurisdiction Study

The experimental design for the national study was based on the study questions identified by the oversight panel, the noted issue-related findings from the literature review, the findings from the jurisdiction interviews, and the initial Phase I Scope of Work (SOW). Following is a list of factors, taken verbatim from the SOW, followed by a summary of the project team recommendation for consideration, which were incorporated in the final evaluation plan:

- Sample size requirements—How many treated sites (number of signalized intersections equipped with red-light cameras) are required?

  Recommendation 1: In the proposed experimental plan, the project provided sample size estimates based on assumptions of crash counts and safety effects. However, the project team believed that it may be more important to ask, "What can be achieved, that is, what effects can be detected and at what significance levels, with the data that are likely to be available?" These estimates were also provided.
- How many other types of sites (comparison or reference sites) are required to distinguish between what is the effect of the treatment and what is the effect of other factors such as changes in traffic flow,

weather, and police reporting practices that may have also changed from the "before" to the "after" period? In other words, what technique will be used to estimate the number, severity, and types of crashes that could have occurred without the improvement?

Recommendation 2: The multivariate empirical Bayes (EB) procedure was proposed for accounting for effects due to regression to the mean, traffic volume changes, and changes in other factors during the analysis period [4]. In this methodology, the comparison group is used as a reference group for estimating annual adjustments for the safety performance functions that are key to the method. The number of crashes in the reference group needs to be large enough for this purpose. It was estimated, based on experience and on guidelines in Harwood et al. (2000), that 100 crashes per year for each crash type would be required [24].

- Spillover effect—There may be a large spillover effect of camera enforcement to intersections in the same community that are not equipped with cameras. The automated enforcement may provide general deterrence against red-light violations and crashes with effects not limited to specific intersections with cameras.

Recommendation 3: The project team proposed that (a) this spillover effect be estimated separately from the specific deterrence effect and that it be related, if possible, to the proximity of a noncamera equipped intersection to an RLC site and (b) that untreated signalized intersections in the same community not be used as comparison sites to control for crash trends between the before and after periods. It was felt that these sites may be from neighboring communities or may be unsignalized intersections in the same jurisdiction. The use of signalized intersections from other communities raised issues of differences in crash-reporting practices between cities so the unsignalized intersection option was preferable. However, it was anticipated that the use of unsignalized intersections as the comparison group would make the study somewhat more difficult due to the fact that while the major cities interviewed tend to have detailed traffic count data on signalized intersections, not all have an abundance of data on unsignalized intersections.

- Types of crashes—It is desirable to obtain data that contain sufficient detail to identify crashes that were specifically red-light-running events. If it is not possible to categorize crashes specifically as red-light-running events, then the contractor should attempt to define red-light-running crashes as best as possible, given the data available. In addition, obtaining statistics for rear end crashes is necessary because they might increase under RLC enforcement due to increased driver compliance with stopping at red lights.

Recommendation 4: Given the quality of available databases and that a retrospective study was required, it was felt that it would be impossible to achieve the desired goal of estimating the effects for crashes that were specifically red-light-running events. However, the project team believed that there are reasonable surrogates and proposed that, as a minimum, effects be estimated for left-turn opposed, right angle from adjacent approaches and, if possible, crashes in which a driver was charged for a red-light violation. Effects so estimated could be deemed to be conservative. It was also proposed that the effects on rear end crashes be separately estimated and that an assessment be undertaken of the net effects on crashes by considering the relative severities of the different crash types affected.

- Selection bias—The accident history of an intersection during the "before" period is an important clue to what would have been its safety performance during the "after" period. However, that same accident history may also be one of the reasons that that particular intersection was selected for treatment. This factor makes prediction of safety performance during the "after" period subject to bias called "regression to the mean" (RTM). In addition, the RTM bias may be present even if locations were not selected because of an abnormally high or unusually low rate of crashes.

Recommendation 5: Substantial evidence of this bias was found in previous evaluations. The interviews also substantiated that virtually all cities use "high crash counts" as a factor in choosing the RLC locations (as would be expected and logical). Therefore it was deemed critical that regression to the mean be properly accounted for using the most advanced state of the art methodology. To this end, in Recommendation (2), the EB methodology was proposed; this methodology has other attractive features, e.g., the fact that it mitigates the difficulties in more traditional methods of identifying a comparison group for accounting for regression to the mean.

- Evaluation periods—The adoption of RLC enforcement is relatively recent in the United States; therefore, "after" periods will be short in most cases. The contractor should design a research study that will obtain as much "after" data as possible.

  Recommendation 6: Naturally, the research study should be designed to obtain as much "after" period data as possible. At the same time the study should ensure that the after period is at least one year long and that it ends at a time when it is certain that all crashes occurring have entered the database. On a related issue, the ability to detect an effect also depends on the before period crash count. Fortunately, the EB methodology is such that much longer before periods can be used than the period of two to four years employed in conventional before-and-after comparisons. Unfortunately, the available "before" data in cities is not as extensive as would be the case for State-controlled rural roads. Of the 15 cities interviewed, two have virtually no before-period data, three have 2 years, five have 3–4 years, and three have more than 5 years. The remaining two did not know how many years of data were available. Thus, two-thirds have 3 years or more.

The remainder of this section presents details of the design on the basis of these recommendations and data available in the jurisdictions interviewed.

Study Design Details

A detailed study design was proposed to the project oversight panel. The panel approved this plan after limited modification. The following text describes the design as ultimately implemented by the project team.

Basic Objectives and Main Analytical Requirements

The basic objective was to estimate the change in target crashes. Following is a list of possible target crash types:

- Right-angle (side impact).
- Left-turn (two vehicles turning).
- Left-turn (one vehicle oncoming).
- Rear end (straight ahead).

- Rear end (while turning).
- Other such as crashes specifically identified as red-light-running.

These were estimated separately for two groups of sites:

- Sites where cameras are used (specific deterrence effects).
- Signalized intersections without cameras at various levels of proximity to the treatment sites.

The preparation of a study design entailed both the preparation of a data collection plan and an analysis plan. The analytical requirements to provide the desired estimates drove the data collection needs.

The analysis examined the safety effect of red-light-camera enforcement to provide insights into a number of issues, within the confines of available data. The data collection plan, discussed later in this report, provided insights into the capacity of the available data to address these issues.

Meeting the objectives and addressing the key issues placed the following list of special requirements on the data collection and analysis tasks:

- Select a large enough sample size to detect, with statistical significance, what may be small changes in safety for some crash types.
- Select carefully comparison sites or cities to ensure that safety at these sites is unlikely to be affected by the RLC installations.
- Account properly for traffic volume changes.
- Pool data from several jurisdictions to improve reliability of results and facilitate broader applicability of research products.

Overview of the General Evaluation Methodology

The general analysis methodology used is different from those used in the past, benefiting from significant advances made in the past 10 years in the methodology for the conduct of observational before-and-after studies, which culminated in a landmark book by Hauer [4]. That book also provides guidance on study design elements such as size and selection criteria for treatment and comparison groups and the pooling of data from diverse sources. All these are crucial elements in successfully conducting a study to obtain results that will have wide applicability.

The evaluation considered the issues identified earlier on the basis of panel input and the literature review and survey to the extent that is practical. The inclusion of a variable in the analysis was ultimately resolved on the basis of whether relevant data could be obtained within the confines of the project, and whether obtainable sample sizes and the variation in levels of a variable were sufficiently large to isolate its effects (if any).

The methodologies documented by Hauer range from simple before-and-after comparisons to the more powerful EB methodology [4]. The team proposed that the latter approach be pursued in seeking to overcome the difficulties associated with conventional before-and-after comparisons, while providing a fresh approach to overcome the limitations of previous evaluations of red-light cameras. Specifically, the analysis would:

- Properly account for regression to the mean.
- Overcome the difficulties of using crash rates in normalizing for volume differences between the before-and-after periods.
- Reduce the level of uncertainty in the estimates of safety effect.
- Provide a foundation for developing guidelines for estimating the likely safety consequences of contemplated RLC installation.
- Properly account for differences in crash experience and reporting practice in amalgamating data and results from diverse jurisdictions.
- Avoid the difficulties of conventional treatment-comparison experimental designs caused by possible spillover and/or migration effects to natural comparison groups.

In the EB approach, the change in safety for a given crash type at an RLC intersection is given by equation 1:

$$\pi - \lambda \tag{1}$$

where π is the expected number of crashes that would have occurred in the after period without the cameras and λ is the number of reported crashes in the after period.

In estimating π, the effects of regression to the mean and changes in traffic volume were explicitly accounted for using safety performance functions (SPFs) relating crashes of different types and severities to traffic flow and other relevant factors for each jurisdiction based on a reference group of signalized intersections without RLCs. Annual SPF multipliers were calibrated to account for the temporal effects on safety of variation in weather, demography, crash reporting and so on. Because of the possibility of spillover effects to the reference group of signalized intersections, it was decided to estimate the annual multipliers for the period after the first RLC installation from the trend in annual multipliers of SPFs calibrated for a comparison group consisting of unsignalized intersections in the jurisdiction.

In estimating the SPFs a parameter *k*, which is a constant for a given model, is iteratively estimated with the use of a maximum likelihood procedure. (In that process, a negative binomial distributed error structure is assumed with *k* being the dispersion parameter of this distribution; the estimated value of *k* is the one that maximizes the likelihood of observing the crash counts, given the calibrated SPF.) See Hauer for more detail [4].

Empirical Bayes before-and-after Evaluation Example

An illustration of the EB before-and-after evaluation methodology is provided next. Full theoretical details can be found in Hauer [4]. This example of the evaluation methodology is applied at a site and aggregate level. Note that the data presented are for illustrative purposes only, and do not represent data collected in this study.

Data and SPFs

Consider an intersection at which RLC was implemented in September 2000. Refer to this as site (i). Suppose it is desired to estimate the effect of RLC on right angle injury crashes at this site.

Suppose the SPF for right angle injury crashes for a given year *y* in the before period for this jurisdiction is:

$$Injury\ crashes/year = \alpha_y\ (MAJAADT_y)^{0.400}(MINAADT_y)^{0.811} \quad (2)$$

where α_y is the calibrated multiplier for this jurisdiction for a given year using the recalibration procedure, and $MAJAADT_y$ and $MINAADT_y$ are respectively the major and minor road entering average annual daily traffic (AADT) in year *y*.

For this illustrative example, it is assumed that the recalibration process has been completed, and that the values of α_y are as given in table 6. That process also calibrated a value of 1.44 for the negative binomial distribution overdispersion parameter *k* that is used in the EB procedure. Crashes are available from 1996–2001 as shown in table 2. For illustrative purposes, assume that the yearly entering AADTs were either all available or some were estimated using a procedure in Lord [25]. The AADTs are also shown in table 2.

Effect for Site (i)

The calculations in table 2 are based on the methodology in Hauer [4]. They pertain to a single site (i) for which the results show that $\pi(i) = 4.384$ right angle injury crashes are expected in the after-period without treatment. Four such crashes ($\lambda(i)$) were recorded.

The crash modification factor for this one site is, from Hauer: [4]

$$\theta = [\lambda(i)/\pi(i)]/[1+ \mathrm{Var}\{\pi(i)\}/ (\pi(i))^2] = [4/4.384]/[1 + 0.820/4.384^2] = 0.875 \quad (3)$$

It means that the point estimate of the crash reduction is 100(1- 0.875) = 12.5%.

Table 2 shows the AADTs for the EB evaluation example.

Table 2. Summary of results for right-angle injury crashes at site (i)

	1996	**1997**	**1998**	**1999**	**Jan-Aug 2000**	**Oct-Dec 2000**	**2001**
Crashes in year(X)	4	6	3	5	4	1	3
	Sum = **Xb** = 22					Sum = λ = 4	
MAJAADT	41302	42169	43460	43891	44321	42322	42875
MINAADT	3596	3671	3783	3821	3858	3720	3520
Model Alpha × 10-5	1.32	1.45	1.20	1.25	1.38	1.38	1.35
Parameter k	1.44	1.44	1.44	1.44	1.44	1.44	1.44
Model Prediction $E\{\kappa_{i,y}\}$	0.709	0.799	0.686	0.723	0.538	0.192	0.724
$C_{i,y} = E\{\kappa_{i,y}\}/ E\{\kappa_{i,1}\}$	1	1.126	0.967	1.019	0.759	0.271	1.020

The standard deviation of θ (from Hauer) is given by:(4)

$$Var(\theta) = [\theta^2\{(1/\lambda(i)) + [\mathrm{Var}\{\pi(i)\}/ (\pi(i))^2]\} / \{1+[\mathrm{Var}\{\pi(i)\}/\pi(i)^2]\}^2\}]^{0.5}$$

$$= [0.875^2 \{(1/4)+[0.820/4.384^2]\}/\{1+[0.802/4.384^2]\}^2]^{0.5} = 0.453 \quad (4)$$

Aggregate Effects Over Several Sites

Results for a single site will tend to be meaningless and almost certainly lack reasonable statistical significance. To aggregate results over several sites, the procedure is to simply add their individual values of the λ(i), π(i), and Var{π(i)}over all sites and replace these values in the equations with their respective sums, λ_{sum}, π_{sum}, and Var{π_{sum})}.

For illustration, five sites with various lengths of before and after periods are added to the analysis. Assume that the observed number of crashes, λ(i), the expected number of crashes without treatment, π(i), and its variance Var{π(i)} have been calculated and are given in table 3. In this example one site (Site 4) experienced more crashes in the after period than expected.

The crash modification factor for the five sites is, therefore:

$$\theta = (\lambda_{sum}/\pi_{sum})/\{1 + [Var(\pi_{sum})/\pi_{sum}^2]\} = [38/46.052]/[1 + 7.260/46.052^2] = 0.822 \quad (5)$$

This means that the point estimate of the composite crash reduction percentage of 100(1-0.822) = 17.8 percent.

The standard deviation of this composite θ is:

$$Stddev(\theta) = [\theta^2\{[Var(\lambda_{sum})/\lambda_{sum}^2] + [Var(\pi_{sum})/\pi_{sum}^2]\}/[1 + Var(\lambda_{sum})/\lambda_{sum}^2]^2]^{0.5}$$

$$= [0.822^2\ \{(1/38)+[7.260/46.052^2]\}/\{1+[7.260/46.052^2]\}^2]^{0.5} = 0.0140 \quad (6)$$

By including more sites in the analysis, some with relatively long after periods, θ has been estimated more accurately with a standard deviation of 0.140 compared to 0.453 using only one site with a short after period. Table 3 shows the composite effect at five sites, for illustration.

Table 3. The composite effect over several sites (for illustration)

Site Number, i	λ (i)	π(i)	Var{π(i)}
1	4	4.302	0.802
2	5	5.555	1.033
3	10	13.250	2.065
4	5	4.500	0.820
5	14	18.450	2.540
Sum	38	46.052	7.260

Data Collection Plan

Based on the requirements for the methodology described, the project team then developed a data collection plan. The first aspect of the plan was the choice of jurisdictions to include in the study. In the following description, this choice was based on sample size needs and the data available in each jurisdiction. The second aspect of the plan involved the specification of data variables to be collected.

Choice of Jurisdictions

Evaluation Study Sample Size Estimation

When planning a before-and-after safety evaluation study it is vital to ensure that enough data are included such that the expected change in safety can be statistically detected. Even though in the planning stage the expected change in safety is not known, it is still possible to make a rough determination of how many sites are required based on the best available information about the expected change in safety. Alternatively, one could estimate, for the number of available sites, the change in safety that can be statistically detected. For a detailed explanation of sample size considerations, as well as estimation methods, see chapter 9 of Hauer [4]. The sample size analysis presented in this section addressed two cases: 1) how large a sample is required to detect statistically an expected change in safety and 2) what changes in safety can be detected with likely available sample sizes. The focus is on detecting effects at the treatment sites. It was assumed (and was later verified) that the number of noncamera equipped signalized intersections will be sufficiently large that spillover effects, if present, will be detected.

Case 1—Sample Size Required to Detect an Expected Change in Safety

For this analysis, it was assumed that a conventional before-and-after study with comparison group design would be used, because available sample size estimation methods are based on this assumption. The sample sizes estimates so provided would be conservative in that EB methodology proposed would require fewer sites. To facilitate the analysis, it was also assumed that the number of signalized reference sites is equal to the number of treatment sites. This assumption was very conservative, because it was later decided to attempt to collect data on three signalized reference sites for each treatment site to better explore the spillover effect.

The statistical accuracy attainable for a given sample size is described by the standard deviations of the estimated percentage change in safety. From this, one can estimate P-values for various sample sizes and expected change in safety for a given crash history. A set of such calculations is shown in table 8 based on assumptions of 20 crashes/site-year of which 3.5 are right angle crashes and 12.0 are rear end crashes, 3 years of "before" crash counts, 1.5 years of "after" period crash counts. The crash rates are estimated as an average of published data for RLC sites in Charlotte, NC, and Howard County, MD. The calculations are based on methodology in Hauer and a spreadsheet on his Web site, http://www.roadsafetyresearch.com [4].

Table 4. P-values for various sample sizes and expected changes in safety*

Number of treated sites	20			60			100		
Percentage change in crashes	10	20	30	10	20	30	10	20	30
Right-angle crashes	0.58	0.23	0.05	0.42	0.08	<0.01	0.33	0.05	<0.01
Rear end crashes	0.51	0.23	0.10	0.42	0.14	0.04	0.40	0.12	0.03

*Based on 3.5 right angle and 12.0 rear end crashes/site-year, and before and after periods of 3 and 1.5 years respectively.

The shaded cells in table 8 indicate where P-values of at least 0.10 are attainable. Thus, for example, if the sample contains 20 treated sites, and a 30-percent reduction in the number of right-angle crashes is expected because of the RLC installation, one may expect to obtain a statistically significant result at the 10 percent level (P = 0.05). With 60 treated sites, if there is a 20-percent increase in the number of rear end crashes, one may not expect a statistically significant result at the 5 percent level (P > 0.10); however, that result would be significant at the 15 percent level.

Case 2— Safety Change Detectable with Likely Available Sample Sizes

On the basis of preliminary crash data available early in the study, an estimate was made of the maximum percentage change in crash frequency that could be statistically detectable at 5-percent and 10-percent significance levels.

Estimates were prepared for a variety of severity and impact types and for four representative jurisdictions from the survey. Because it was likely that additional data would become available as the study proceeded, it was felt that these estimates could be regarded as conservative. The estimates were also conservative based on other considerations mentioned below.

The crash rate assumptions in table 5 were used for this exercise. They were based on published data from RLC installation sites in Howard County, MD (HC), and Charlotte, NC (CH), and on typical severity ratios indicating that about 35 percent of all crashes at signalized intersections involve injuries.

Table 5. After period crash rate assumptions

Crash Type	Rate A (crashes/intersection/year) (HC)	Rate B (crashes/intersection/year) (CH)
All right-angle	4.5	2.4
Injury right-angle	1.6	0.9
All rear end	20.7	6.4
Injury rear end	7.3	2.3

Tables 6 and 7 show, for each crash type, an estimate of the number of intersection years of after-period crash data for treatment sites in each of several jurisdictions and for 2 groups of jurisdictions. Separate parts of the table are presented for right-angle and rear end crash effects. For the two crash rate assumptions it shows the maximum percentage change in crash frequency that would be statistically detectable at 5 percent and 10 percent levels of significance for both crash rate assumptions. (For Howard County and Charlotte, only their respective crash rates from table 6 are used.) Only jurisdictions with 10 or more intersection years in the after period were considered as feasible for this analysis. Assuming that some sort of a national estimate would be useful and could be obtained through amalgamation of the results over several jurisdictions, which is certainly possible for injury crashes at least, calculations were also shown for two groups of jurisdictions:

- All jurisdictions listed
- The "top seven" which includes jurisdictions for which significant effects are likely for all crash severities (Howard County, Baltimore, Charlotte, San Diego, San Francisco, Montgomery County, and El Cajon City) but excludes New York City because its data were found

to be unsuitable for use in this study. These “top seven” were ultimately included in the study.

This presentation allowed for various options for deciding on the size of the planned retrospective study; nevertheless, it was felt that consideration must also be given to the ease or difficulty of obtaining quality data from each jurisdiction (which is why New York City was ultimately excluded).

Table 6. Sample analysis for right-angle crash effects

Crash type	Jurisdiction	Intersection-years in after period (and # camera sites)	Minimum percentage of change* detectable for two crash-rate assumptions (A, B)	
			P = 0.10	P = 0.05
All right-angle	New York City, NY	126 (30)	18,21	21,24
	Howard Co., MD	88 (26)	20	22
	Baltimore, MD	85 (48)	20,23	23,26
	Charlotte, NC	69 (30)	24	28
	San Diego, CA	43 (19)	23,28	27,32
	San Francisco, CA	37 (17)	24,29	28,33
	Montgomery Co., MD	35 (15)	25,30	28,34
	El Cajon City, CA	35 (9)	25,30	28,34
	Sacramento, CA	28 (18)	26,32	30,36
	Prince George’s Co.,	20 (16)	29,35	33,39
	Arlington, VA	15 (5)	32,38	36,43
	Chandler, AZ	12 (8)	34,41	38,46
	Boulder, CO	10 (3)	36,43	40,48
	Group 1[a]	392 (104)	16,17	19,19
	All	603 (224)	15,16	18,18
Injury right-angle	New York City, NY	126 (30)	23,27	26,31
	Howard Co., MD	88 (26)	25	29
	Baltimore, MD	85 (48)	26,30	29,34
	Charlotte, NC	69 (30)	32	37
	San Diego, CA	43 (19)	31,37	36,42

Crash type	Jurisdiction	Intersection-years in after period (and # camera sites)	Minimum percentage of change* detectable for two crash-rate assumptions (A, B)	
			P = 0.10	P = 0.05
	San Francisco, CA	37 (17)	33,39	37,44
	Montgomery Co., MD	35 (15)	33,40	38,45
	El Cajon City, CA	35 (9)	33,40	38,45
	Sacramento, CA	28 (18)	36,43	40,47
	Prince George's Co.,	20 (16)	40,47	44,52
	Arlington, VA	15 (5)	43,51	48,56
	Chandler, AZ	12 (8)	46,54	51,59
	Boulder, CO	10 (3)	48,56	53,61
	Group 1[a]	392 (104)	18,20	21,23
	All	603 (224)	17,18	20,21

* Assumes a decrease in right-angle crashes and an increase in rear end crashes.

[a] Group 1 includes Howard County, Baltimore, Charlotte, San Diego, San Francisco, Montgomery County and El Cajon City.

Table 7. Sample analysis for rear end crash effects

Crash type	Jurisdiction	Intersection-years in after period (and # camera sites)	Minimum percentage of change* detectable for two crash-rate assumptions (A, B)	
			P = 0.10	P = 0.05
All rear end	New York City, NY	126 (30)	22,25	27,32
	Howard Co., MD	88 (26)	22	28
	Baltimore, MD	85 (48)	23,28	28,35
	Charlotte, NC	69 (30)	27	37
	San Diego, CA	43 (19)	25,34	31,43
	San Francisco, CA	37 (17)	25,36	32,46
	El Cajon City, CA	35 (9)	26,37	32,47
	Sacramento, CA	28 (18)	27,41	34,52
	Prince George's Co.,	20 (16)	30,48	37,62
	Arlington, VA	15 (5)	33,56	41,74

Table 7. (Continued)

Crash type	Jurisdiction	Intersection-years in after period (and # camera sites)	Minimum percentage of change* detectable for two crash-rate assumptions (A, B)	
			P = 0.10	P = 0.05
	Boulder, CO	10 (3)	38,73	49,99
	Montgomery Co., MD	35 (15)	26,37	32,47
	Chandler, AZ	12 (8)	35,65	45,87
	Group 1a	392 (104)	21,21	26,26
	All	603 (224)	20,21	25,26
Injury rear end	New York City, NY	126 (30)	25,33	31,42
	Howard Co., MD	88 (26)	26	33
	Baltimore, MD	85 (48)	27,39	33,50
	Charlotte, NC	69 (30)	43	56
	San Diego, CA	43 (19)	32,55	41,73
	San Francisco, CA	37 (17)	34,60	44,81
	Montgomery Co., MD	35 (15)	35,63	45,84
	El Cajon City, CA	35 (9)	35,63	45,84
	Sacramento, CA	28 (18)	38,73	49,99
	Prince George's Co.,	20 (16)	45,93	58, >100
	Arlington, VA	15 (5)	52, >100	68, >100
	Chandler, AZ	12 (8)	59, >100	79, >100
	Boulder, CO	10 (3)	67, >100	90, >100
	Group 1a	392 (104)	22,23	27,29
	All	603 (224)	21,22	22,28

* Assumes a decrease in right-angle crashes and an increase in rear end crashes.

[1a] Group 1 includes Howard County, Baltimore, Charlotte, San Diego, San Francisco, Montgomery County and El Cajon City.

Sample Design Conclusions

Judgments on the likelihood of detecting significant effects assume that there is, in fact, an effect on crashes. If an effect does not exist, of course, no effect will be statistically detectable. Table 8 presents the authors' best judgment during the sample design stage on the likelihood of detecting (at the 10-percent level) safety effects expected on the basis of the literature review, which revealed that it is not unreasonable to expect effects on the order of a 25-percent decrease in right-angle crashes and a 30-percent increase in rear-end crashes. Even so, for reasons explained earlier, these judgments were based on results that are likely to be conservative.

Table 8. Best judgment on possibility of detecting safety effects

	All right-angle	Injury right-angle	All rear end	Injury rear end
New York City, NY	√	√	√	√
Howard Co., MD	√	√	√	√
Baltimore, MD	√	√	√	√
Charlotte, NC	√	X	√	X
San Diego, CA	√	X	√	X
San Francisco, CA	√	X	√	X
Montgomery Co., MD	√	X	√	X
El Cajon City, CA	√	X	√	X
Sacramento, CA	X	X	√	X
Prince George's Co., MD	X	X	√	X
Arlington, VA	X	X	X	X
Chandler, AZ	X	X	X	X
Boulder, CO	X	X	X	X
Group 1	√	√	√	√
All	√	√	√	√

√ = significant results may be obtained
X = significant results may not be obtained.

Selection of Study Jurisdictions

As noted, the Us in table 8 were based on one criteria for inclusion in a study of RLC effects— the available crash data. However, there were other criteria that needed to be considered—the availability and quality of the other data. Table 9 summarizes these crash-related findings from table 8, along with information on other data extracted from the interview forms. While a variety of information was captured in the interview, because of both study needs such as traffic flow data and the high-priority questions of interest, emphasis was placed on the presence of data on yellow interval changes, traffic flows at signalized and unsignalized intersections, the level of publicity campaign (to attempt to get a range of levels), and the type of signing. Note that the level of the publicity campaign was a project staff judgment based on the jurisdiction's response to the initial telephone questionnaire discussion concerning public information. In the questionnaire, the three levels were given the following definitions:

- *High*—A major planned public information campaign including components such as the FHWA public information program, combined PI efforts with other departments in the jurisdiction (e.g., local health department), and television spots.
- *Medium*—Moderate public information program with limited expenditures, but good coverage of the RLC effort by news media.
- *Limited*—Limited public information program with media coverage only from interviews or press conferences, or both.

Table 9 shows 3 categories of the authors' judgment of the best cities based on these data variables. Howard County, MD, was judged the best overall, shown with bold italics. The second group of cities (in some order of preference) are those in italics—Baltimore, MD, Charlotte, NC, San Diego, CA, San Francisco, CA, Montgomery County, MD, and El Cajon City, CA. Each has some shortcomings, either in crash sample size or other data. For example, Baltimore has limited traffic count data, and El Cajon City's crash data require further investigation. The remaining cities had serious problems either in crash counts or other data or in the size of the sample.

Table 9. Best judgment on sites to use based on crash and non-crash data available

	Significant crash types	Signal data available	Traffic flow data	Publicity campaign
New York City, NY	All 4 types	YI, CL, SC	None	HPI, none
Howard Co., MD	***All 4 types***	***YI, ARI, CL, SC***	***FTC, UITC***	***MPI, SO***
Baltimore, MD	*All 4 types*	*YI, ARI, CL, SC*	*LTC*	*MPI, SI*
Charlotte, NC	*ARA, ARE*	*YI, ARI, CL, SC*	*FTC, UITC*	*HPI, SB*
San Diego, CA	*ARA, ARE*	*YI, ARI, CL,*	*FTC, UITC*	*HPI, SI*
San Francisco, CA	*ARA, ARE*	*YI, ARI, CL, SC*	*LTC, UITC*	*HPI, SB*
Montgomery Co, MD	*ARA, ARE*	*YI, ARI, CL, SC*	*LTC, UITC*	*HPI, SO*
El Cajon City, CA	*ARA, ARE*	*YI, ARI, CL, SC*	*FTC, UITC*	*MPI, SB*

	Significant crash types	Signal data available	Traffic flow data	Publicity campaign
Sacramento, CA	ARE	YI, ARI, CL, SC	FTC, UITC	HPI, SB
Prince George's Co., MD	ARE	YI, ARI, CL, SC	LTC, UITC	LPI, SO
Arlington, VA	None	YI, ARI, CL, SC	None	MPI, SO
Chandler, AZ	None	YI, ARI, CL, SC	FTC, UITC	MPI, SO
Boulder, CO	None	YI, ARI, CL, SC	FTC	HPI, SO
Fairfax Co., VA	No after data	ARI, CL, SC	Unclear	MPI, SI
Greensboro, NC	No after data	ARI, SC	FTC	MPI, SI

Significant Crash Types: ARA—All right-angle; IRA—Injury right-angle; ARE—All rear end; IRE—Injury rear end

Signal Data Available: YI—Yellow interval (length and changes); ARI—All-red interval (presence and changes); CL—Cycle length, SC—Signal coordination

Traffic Data Available: FTC—Full traffic counts (i.e., regular program of traffic counts for all signalized intersections); LTC— Limited traffic counts (some intersections, or only as requested); No—no traffic counts; UITC—Unsignalized intersection traffic counts

Public Information: HPI—High public information campaign; MPI—Medium public information campaign; LPI—Limited public information campaign; SI—Warning signs at intersections; SO—Warning signs at other locations (e.g., edge of town or corridor); SB—Warning signs at both intersection approaches and other locations

Note that, as indicated earlier, New York City, the site with the largest crash sample, fell into the infeasible group because of the lack of traffic count data. New York City was also considered different from other U.S. cities by an RLR expert outside the project team because of its size, the number of intersections, the small proportion of intersections that are treated, the possible dilution of any publicity campaign, high tourism, and other factors.

As can be seen, three of the cities appeared superior to the other four cities—Howard County and Baltimore, MD, and Charlotte, NC. While a multijurisdictional study could be done with just these three cities, the project team did not recommend that because the sample size of jurisdictions would be small, two are in Maryland, and all three are eastern cities.

Data Collection Requirements

It was recommended that, for any installation a minimum of 2 years of before-period and 1 year of after-period information should be available and that, ideally, at least 3 years of crash data be required for each of these periods. As found later in the actual data collection, all sites in all jurisdictions had 4 to 9 years of before-period data, with an average before-period of 6 years.

The length of the after period data varied from less than 1 year (in approximately 8 percent of the sites) to 5 years, with an average after-period of approximately 2.76 years for all sites.

Table 10. Data items required

Crash variables	Traffic variables	Geometric variables	Operational variables	Other variables
Crash date Crash time Weather condition Light condition Impact type Crash severity Initial direction of vehicles Vehicle maneuver Vehicle type Location type (intersection) Citation issued	Major, minor road entering AADT (at least one count in each of before- and after-period	Number of approach lanes Median presence/width Approaches with left-turn lanes Approaches with right-turn lanes	Cycle lengths Yellow intervals All red intervals Signal coordination Left-turn priority phasing Speed limit	RLC installation date Grace periods Ticketing protocols RLC approach Date, type of other significant changes to intersection RLC signing Presence of LEDs, back plates, and 30.48 cm (12 in) lens for signals (treatment sites only)

Crash data for the same years were also required for a reference group of locations, similar to the RLC locations, except that these were not equipped with RLCs. These sites were to be used in the recalibration of safety performance functions and to investigate possible spillover effects. Because the reference group was to be used both for this SPF recalibration and to study the distance-of-influence issue, a later decision was made to attempt to identify three reference- group signalized intersections for each treated intersection in a jurisdiction. To account for time trends between the before-and after-periods, crash data also were collected from a comparison group of approximately 50 unsignalized intersections in each jurisdiction.

Table 10 lists the basic data items originally proposed for collection. As will be described in the later section concerning data collection, modifications were made to this listing based on available data and funding issues.

VIII. Study Methodology for Economic Analysis

As noted, two experimental plans were developed for this study. The first, described in detail in the preceding sections, was aimed primarily at examining changes in crash frequency resulting from the installation of RLC systems. The second experimental plan involved development of a database and methodology to examine the costs and benefits of RLCs in economic terms, thus allowing changes in crash severity and frequency to be analyzed simultaneously and identification of factors associated with the greatest RLC benefits.

The use of economic analysis in the study of RLCs is important because of the nature of the expected effect of the treatment. Based on past research, RLCs probably will lead to decreases in right-angle crashes, for which injuries are often severe, and increases in rear end crashes, for which injuries tend to be less severe. It is also possible that within each of these crash types, the average level of injury severity is different with and without RLCs; therefore, it is probable that an evaluation of the economic benefits could result in a different conclusion about the effect of RLCs than might result from an analysis of changes in crash frequency by crash severity (e.g., the examination of changes in the frequency of injury crashes). Gaining a handle on the economic benefits is also important from the perspective of being able to assess cost- effectiveness of RLCs and to compare this against the cost-effectiveness of other safety treatments aimed at reducing crashes at signalized intersections.

To combine changes in both crash frequency and injury severity, each crash can be characterized by one measure—a dollar value based on the average level of injury severity for that type of crash. After that conversion is done, the results of the analysis can be reported in terms of changes in total economic cost of crashes expected without RLCs and those that actually occurred in the period after RLC installation. The unit crash costs can then be applied to the crash frequencies recorded after RLC implementation and those expected without RLC as estimated by the empirical Bayes method being used for evaluation.

The key is the successful conversion of police-reported crash injury levels to a set of acceptable dollar-cost measures. As described in the following sections, the team explored different alternatives for this economic conversion and then worked with FHWA to define a final study design.

Initial Economic Analysis Study Design

In Charlotte, NC, there were accessible files of hospital and emergency room data that could be linked to traffic crashes, the initial discussion of an economic analysis study design centered on using these linked data. After further thought and exploration, it was clear that the possible use of these data raised a number of critical issues:

- The available database covers only one of the two major hospitals in the catchment area, meaning that information for the remaining patients in the other hospitals would need to be estimated. That is, it must be assumed that there are no shifts from one hospital to the other between the before-and after-period, and that the medical cost changes seen in one hospital would be similar in the other hospital.)
- The database would not include information on nonhospitalized fatalities or on nonhospitalized occupants experiencing either lesser injuries or no injuries. Because the proportion of car crash occupants who are hospitalized is a small proportion of the total crash-involved population, costs for the nonhospitalized groups would need to be developed from other sources.
- Most important, the economic data available in the database to be linked with crashes would contain only medical/hospital costs, which is only one component of either human capital cost or comprehensive cost.

FHWA agreed that the issues were significant, and it required that a feasibility study be conducted to determine if these issues could be overcome.

The following section describes the results of this exploration.

Examination of Alternative Economic Analysis Methods

The research team identified the following major issues that require resolution to successfully conduct an economic study:

- RLC systems potentially could affect the full injury distribution from fatally injured to no injury; thus, the economic cost chosen must cover this complete distribution.
- Hospital-related data (even that including costs for emergency room visits in addition to hospital admissions) include perhaps 15 percent of the total crash population. The data do not include information on those who die before reaching the hospital nor those whose injuries do not require hospital or emergency room treatment.
- Medical costs most likely to be captured by hospitals are only one element of the total cost of crashes. Other elements such as lost work productivity, rehabilitation costs, insurance cost, quality of life losses, and others are not captured in medical cost data.
- As requested by FHWA, data on both "comprehensive costs" (including "willingness to pay" costs related to pain and suffering) and "human capital costs"[4] (including many other economic elements, but not "pain and suffering") should be considered in this analysis.
- Research conducted by and for the U.S. Department of Transportation (DOT) National Highway Traffic Safety Administration (NHTSA) has provided a detailed analysis of human capital costs for motor-vehicle crash victims in terms of Year 2000 dollars. Unfortunately,

---

[4] A detailed discussion of "human capital" and "comprehensive" costs can be found in Blincoe, et al. [26] In summary, human capital costs include direct and indirect costs to individuals and society as a whole from the decline in the general health status of those injured in motor vehicle crashes. Components include medical and rehabilitation costs, emergency services costs, lost market productivity, lost household productivity, insurance administration, workplace costs, legal costs, travel delay costs, and property damage costs. Comprehensive costs include all these components plus additional costs associated with intangible consequences of crashes to individuals and families such as pain and suffering and loss of life. In studies of motor vehicle crashes, both types of costs are usually keyed to individual levels of injury severity measured on the AIS within an individual body part (because consequences can vary by severity within body part injured). AIS is specified by trained medical data coders, usually within a hospital context. Average human capital or comprehensive costs are often defined in reports for an eight-point injury scale based on the Maximum AIS score (MAIS) for an individual. Appendix A of Blincoe, et al., shows the average human capital cost ranges from approximately two-thirds of the comprehensive cost for a minor (MAIS 1) injury level to less than one-third the comprehensive cost for a fatality (i.e., $0.98 million versus $3.37 million, in Year 2000 dollars) [26].

these costs are keyed to injury severity levels defined by the Abbreviated Injury Scale (AIS) rather than by the KABCO[5] scale found in police reports, which will be the basic data being provided by the participating localities for this evaluation.

- There has been no conversion from AIS-based cost to KABCO cost for FHWA since 1994. There is agreement that the updating of the 1994 data to current year costs using changes in the Consumer Price Index would not be sufficient because these costs would not be consistent with the current NHTSA research nor current economic cost for fatalities being used by other federal agencies, and the 1994 costs did not include costs for each of the levels of the KABCO scale, only for "injury" [25]. Because RLCs may cause tradeoffs in injury severities within the injury category (e.g., a decrease in serious injuries and an increase in minor injuries), these 1994 costs are not considered sufficient for this study.

Given these issues, the team explored two alternative approaches that led to exploring two databases containing information on the economic costs of motor vehicle crashes.

Analysis of the Economic Benefits of RLCs Using Maryland CODES Data

First, an attempt was made to find hospital and emergency room cost data similar to that originally proposed from Charlotte. Information on NHTSA's Crash Outcome Data Evaluation Systems (CODES) project was examined on the current CODES Web site. There are no CODES projects in either California or North Carolina, but the State of Maryland has one, where three of the test cities are located. E-mail and telephone discussions were held with the database engineer for the National Study Center for Trauma and EMS, University of Maryland—Baltimore, the current Maryland CODES coordination agency. This group has data on all motor vehicle crashes reported to the State of Maryland beginning in the early 1990s, including nonhospitalized injury and property-damage-only (PDO) crashes. For a typical

[5] The KABCO severity scale is used by the investigating police officer on the scene to classify injury severity for occupants with five categories: K, killed; A, disabling injury; B, evident injury; C, possible injury; O, no apparent injury [7]. These definitions may vary slightly for different police agencies.

CODES analysis, the crash data are linked to hospital discharge data or emergency room data, or both, based on a probabilistic match of variables linking the crash to hospital admission or emergency room visit records. Medical cost data can then be extracted for each driver of each vehicle and linked with the crash data. The project team requested data files from the CODES administrator for all linked and unlinked crashes occurring in Howard and Montgomery Counties and Baltimore City for the period of 1994–2001. After discussions with the administrator, the CODES Advisory Committee overseeing the Maryland data granted approval for the request and data were prepared and sent to the project team.

If these Maryland CODES data were used in the economic analysis, they would provide more complete, but similar data to that of the Charlotte source. While slightly more difficult to work with than the Charlotte-linked data, these data include all hospital admissions and emergency room visits (at least, all that can be linked) rather than just those for one hospital, as was the case in Charlotte.

However, the same problems with the Charlotte study were present in the Maryland data. The only cost data available are medical costs. After again considering how other cost items could be added to the data set, such as how to add costs for lost productivity, perhaps from NHTSA data based on the AIS level of injury, the authors concluded that this study approach was not sufficient.

Economic Benefits Analysis of RLCs Using Pacific Institute for Research and Evaluation Comprehensive Cost Data

As noted earlier, the key to this economic analysis effort is to assign a human capital or comprehensive crash cost to each relevant crash in a test city. If accomplished, the economic analysis could be conducted in each of the seven cities rather than in just one, as originally envisioned. Because the city data are based on police-reported KABCO injury scales, this could be done only if AIS-based comprehensive cost could be mapped to the KABCO scale with suitable precision. That mapping became the goal of the efforts.

Dr. Ted Miller of the Pacific Institute for Research and Evaluation (PIRE), a leading U.S. expert on economic costs for motor vehicle crashes, developed the earlier KABCO-mapped comprehensive costs for FHWA. Because of this, the project team initiated a detailed review of his more recent publications related to this topic. Subsequent discussion with Miller indicated that it was feasible to update the mapping. In addition to providing comprehensive costs for this RLC study, the proposed approach would result in KABCO-mapped costs that could be used by FHWA in future evaluations and problem analysis efforts.

In a 1997 article, Miller et al., used National Automotive Sampling System (NASS) Crashworthiness Data System (CDS) files, Fatal Analysis Reporting System (FARS) data, and General Estimates System (GES) data to develop an estimate for the total direct cost (not human capital or comprehensive costs) and years of functioning life lost for occupants involved in 30 different crash geometries (e.g., cross-path crashes at signalized intersections) [27]. While comprehensive costs were not calculated directly, it is noted that "years of functioning life lost" is a measure that can be converted to comprehensive costs, combining human capital direct and indirect costs with pain and suffering costs, as discussed in appendix A of the earlier cited report [25]. Miller's goal was to better identify the highest cost geometries, and thus the ones to best target for treatment efforts. Because direct (and comprehensive) costs can be developed for AIS levels, his approach involved developing such occupant-based costs for each AIS level in each of the 30 different crash geometries. The NASS-CDS files contained the AIS level for each occupant; however, because the FARS and GES data to be used in the national estimates do not include the AIS level, Miller had to map his AIS costs to KABCO levels. Because a KABCO rating from the original investigating police officer was also included for each occupant in the NASS CDS file, the occupant-injury cost of each level on the KABCO scale within each geometry could also be calculated. Because current CDS files do not contain crashes involving nontowaway vehicles or pedestrians, the old NASS files were used to fill in these missing data. Then, using national crash-occupant frequencies within each KABCO level for each of the 30 crash geometries based on GES and FARS data, Miller et al., produced an estimate of the total national direct cost for each geometry [27].

While this current project's goal was not to produce such national estimates of cost by crash geometry, there was the critical need for human capital and comprehensive cost for each KABCO level within certain intersection-related crash geometries (e.g., right-angle crashes at signalized and stop-controlled intersections). Thus, it appeared logical that Miller could use his data to produce such estimates.

Possible Study limitations in Miller's Dataset:

Three critical study limitations based on files available to Miller were identified by the project team before the development of the final data request: (a) sample size limitations in the NASS and CDS crash files to be used, (b) whether available variables in the Miller files would allow further classification of crash geometry by urban and rural crash location, and (c) whether Miller's earlier occupant-based cost findings could be converted to costs per crash within each geometry of interest.

The possible sample size limitations in NASS files containing both AIS and KABCO injury scaling arose from the need to estimate cost for each KABCO level within each of the more than 30 crash geometries, further categories by urban versus rural crash location. Such detailed estimates had never been attempted before, and low sample sizes could result in both unstable and illogical findings. As indicated later, such problems did arise.

Second, Miller's earlier analysis did not separate crash geometries by urban and rural location or by speed limit. Because KABCO levels are much broader than AIS levels, the cost of injury within any KABCO level for a given crash geometry might differ depending on speed limit or urban and rural location. For example, the severity and thus the cost of A-injury angle crashes at rural higher-speed intersections may be greater than A-injury right-angle crashes at urban intersections. Given this fact, it was desirable to further categorize Miller's 30 geometries by either speed limit or urban and rural location.

Unfortunately, examination of documentation for the databases to be used by Miller indicated no urban and rural indicator in one of the critical files; however, speed limit variables were present. The team then used FARS, NASS, GES, and Highway Safety Information System (HSIS) data from two states to compare crash-related speed limits to various urban versus rural designations. There was significant overlap of limits within urban and rural designations in all three files.

Based on the distributions and the need to have sufficient samples sizes in all the subcategories, a decision was made to attempt to categorize cost estimates by locations with speed limits of 72.42 km/h (45 mi/h) and slower versus 80.47 km/h (50 mi/h) and faster.

Third, the primary need in this study (and in future FHWA studies) is cost estimates per crash rather than per occupant. Fortunately, Miller's preliminary investigations indicated that cost per crash could be developed from the existing databases.

Possible Economic Study Limitations from Small Jurisdictional Samples of Crash Data

This discussion is based solely on information available in Miller's databases; however, the choice of what economic analyses to conduct, and thus what specific cost categories to request from Miller, also depended on the nature of the database to be used in this RLC analysis—the sample size of crashes within different KABCO severity levels in the seven jurisdictions. Two significant issues deserve further attention.

First, in the initial analysis planning, the project team developed estimates of sample sizes necessary for analysis. The team then examined the total right-angle and rear end crash estimates and the injury right-angle and rear end crash estimates from each of the seven jurisdictions to determine if individual site-by-site analysis would be feasible. As shown in table 8, while each of the seven jurisdictions should provide adequate data for a "total crash" analysis, an "injury-crash" analysis may yield only statistically significant effects in two of the seven jurisdictions. This finding was based on combining all injury levels (K,A,B,C) in the latter analysis. Even there, five of the seven individual analyses may not yield significant effects.

In this economic analysis, the best methodology would be to assign a human capital or comprehensive cost to each occupant injury in each of the involved vehicles, and sum these to produce a total crash cost as the outcome variable. This requires assigning a cost to each level of injury, then subdividing the "combined injury" category in more detail. While the costs will be multipliers for each injury, and thus the magnitude of cost as the outcome variable will be much larger than the magnitude of individual injury crash counts, there remains the logical question of whether these small samples of individual injury-level crashes will support such an analysis. While the cost numbers will be large enough, the question is whether the sample is large enough to produce a stable measure of cost in the before-and-after periods for an individual site.

The second issue raised by the sample sizes within the RLC database is whether a cost should be assigned to each severity level, including fatalities, or should some levels be combined. The primary issue here is that even in right-angle crashes, fatalities will represent a small proportion of the total injury distribution (perhaps less than 2 percent). The comprehensive and human capital cost assigned to a fatality will always be multiples of cost for even an A-injury. The final analysis showed that the fatal crash costs were 10 to 30 times the A-injury crash costs within the same crash type; thus, one or two fatalities in either the before-or after-period for a given location could greatly affect the cost-related analysis. While this could be legitimate if one could expect repeated similar samples to have the same number of fatalities, this is not likely, particularly in samples this small. (It is also noted that Hall, in crash-cost research done for the State of New Mexico, argues that not only are such fatalities somewhat random in any crash sample, the main factors determining whether an injury is a fatality rather than a severe injury probably are not affected by roadway-related treatments. They are more likely to be related to factors such as occupant age, restraint use, and type and size of vehicles involved. He argues that such small numbers of fatalities should not be allowed to affect decisions on roadway-based treatments such as RLCs) [28].

The issue becomes whether to assign separate cost to a fatality and an A-injury or combine these two into one category (perhaps representing approximately 10 percent of the injury distribution in these crashes) and assign some type of "average K/A cost."

This same argument can be carried further, combining other categories of injury. Ultimately, this could result in an "average crash cost" (in a similar proportion-weighted manner) for each crash geometry. The cost would be assigned to a crash based on crash type rather than crash injury; however, this would assume that only the crash type frequencies (or ratios) change between before-and after- periods in the RLC evaluation, and that there is no internal shift in injury distribution within a given crash type. Right-angle crashes could be reduced, but it would be assumed that the severity distribution of the right-angle crash in the before-period is the same distribution as in the after-period. While this might be a good assumption based on a hypothesis that RLCs should not change the impact speed in right-angle crashes (because impact speed is a major predictor of crash severity), whether or not this is a good assumption is unknown. It can also be argued that red-light-running right-angle crashes targeted by RLCs tend to be at higher speeds than other right-angle crashes.

A final problem is that the reporting thresholds appear to differ across the seven jurisdictions. For example, PDO crashes are less likely to be reported in California sites and Howard County and Montgomery County, MD, than in Baltimore, MD, or Charlotte, NC. Given that PDOs are assigned a cost in this cost analysis, this differing threshold will lead to different economic outcomes across sites. The same will be true for the frequency analysis of total crashes. This factor will be taken into consideration when the results from different jurisdictions are compared, and when combination of the results is attempted.

Final Methodology for Economic Analysis

Types and Levels of Economic Cost Estimates

A number of issues are present in an attempt to measure the economic effect of RLC systems. Some occur because of a lack of current data on the human capital or comprehensive cost of a crash referenced to individual levels of the police-reported KABCO scale. Others occur because of the nature of the seven-jurisdiction database that will be used in this overall evaluation of RLCs, with available sample sizes of observations in each severity level for pertinent crash types (right-angle and rear end) being the most critical one. Given all these issues and uncertainties, and the fact that the same issues will arise in future FHWA studies, a decision was made to have Miller and the PIRE staff develop multiple levels of both comprehensive and human capital costs. The following levels were requested:

- Level 1—For each of the 22 crash geometries (categorized by two speed limit categories as a surrogate for urban/rural), estimates of cost for crash severity levels K, A, B+C, and O. (Sample size issues in the cost databases made it impossible to develop reasonable estimates of B versus C separately.) This analysis was first done for each of the two speed limit categories, and then with all speed limits combined.
- Level 2—For each crash geometry, estimates of cost when K and A are combined into one cost level and B and C are combined into one cost level – thus K+A, B+C, O. (Again, estimates were calculated with and without categorization by the two speed limit categories.)
- Level 3—Allows for comparison of "injury" versus "noninjury" crashes. Some crash forms (and some reporting officers) define a "C-injury" as a "minor injury" while others define it as a "possible injury." Thus, two definitions of Level 3 costs were used:
  - Level 3A—For each crash geometry (with and without speed limit categorization), estimates of cost when all injuries are combined into one cost level separated from the PDO cost level, thus K+A+B+C versus O.
  - Level 3B—For each crash geometry (with and without speed limit categorization), estimates of cost when K, A, and B injuries are combined into one cost level separated from the C and PDO cost levels, thus K+A+B versus C+O.

- Level 4—For each crash geometry (with and without speed limit categorization), estimates of crash cost without regard to crash severity, in other words, no division by levels of severity.
- Level 5—For each level of crash severity (with and without speed limit categorization), estimates of cost without regard to crash geometry.
- Level 6—Level 5 cost estimates, but with the following categories: K+A, K+A+B, K+A+B+C, B+C, C+O.

Choice of Cost Levels

Because it was not feasible or necessarily desirable to conduct economic analyses at all these levels for each of the seven jurisdictions, a pilot economic analysis study using one jurisdiction was chosen. Charlotte, NC, data were chosen because Charlotte has a medium sample size that provides some knowledge of whether smaller and larger samples can be analyzed, and Charlotte data contain all KABCO crash levels, allowing a multilevel analysis. Here, four economic analyses were conducted, one for each of levels 1–4:

- All KABCO levels within each geometry.
- Costs for K+A, B+C, 0 within each geometry.
- Costs for injury versus no-injury within each geometry (based on both definitions of "no injury").
- "Urban costs" (those with lower speed limits) based on crash geometry only.

The outcomes of these four analyses were used to determine which level is most appropriate given the available sample sizes in the remaining six jurisdictions.

Economic Analysis Statistical Method

The general analysis methodology used to define the economic effects of the RLC program for a given jurisdiction closely parallels the methodology used for total crashes and injury-crash frequencies. Here, instead of the difference between the crashes "expected without treatment" versus "observed with treatment" in the after-period, the measure of effectiveness would be the difference between the net economic costs "expected without treatment" and "observed with treatment" in the after-period.

For simplicity, the theory is presented for estimating the change in crash costs over all treatment sites in a jurisdiction, for a specific crash type, aggregated over all KABCO subgroups (e.g., two subgroups K+A+B+C, O). The crash types of interest are right-angle, rear end, and other (i.e. other than rear end and right-angle). The following notation is used:

CostBi equals the cost of crashes in KABCO subgroup *i* actually occurring at the treatment sites in the before-period.

Λ_{costA} equals the cost of crashes actually occurring at the treatment sites in the jurisdiction in the after-period.

$VAR\{Cost_B\}$ equals the variance of the cost of crashes in the before-period.

$VAR\{\Lambda cost_A\}$ equals the variance of the cost of crashes in the after-period.

Π_{costA} equals the expected cost of crashes in the after-period over all treatment sites had there been no RLC (after correcting for regression to the mean and traffic volume and other differences between before-and after-periods).

$VAR\{\Pi_{costA}\}$ equals the variance of the expected cost of crashes over all treatment sites in the after period without RLC.

B_i equals the observed number of crashes in KABCO subgroup *i* over all treatment sites in the before-period.

Π_i equals the expected number of crashes in KABCO subgroup *i* over all treatment sites in the after-period without RLC (after correcting for regression to the mean and traffic volume and other differences between before-and after-periods). These were derived for the crash frequency analysis presented by Persaud et. al using the empirical Bayes methodology [2].

The estimated change in crash costs is

$$\Phi_{cost} = \Pi_{costA} - \Lambda_{costA}. \tag{7}$$

The variance of change in crash costs is

$$Var\{\Phi_{cost}\} = Var\{\Pi_{costA}\} + Var\{\Lambda_{costA}\}. \tag{8}$$

The cost modification factor is

$$\theta_{cost} = (\Lambda_{costA}/\Pi_{costA})/[1+(VAR\{\Pi_{costA}\}/\Pi_{costA}^2)]. \tag{9}$$

The variance of cost modification factor is given by

$$VAR\{\theta_{cost}\} = \theta_{cost}^2\{[VAR(\Lambda_{costA})/\Lambda_{costA}^2] + VAR(\Pi_{costA})/\Pi_{costA}^2]\}/ [1+(VAR\{\Pi_{costA}\}/\Pi_{costA}^2)]^2. \tag{10}$$

Of interest at this point is how estimates were obtained for the four terms, Λ_{costA}, VAR$\{\Lambda c_{ostA}\}$, Π_{CostA}, and VAR$\{\Pi c_{ostA}\}$. Following are the approximate methods used.

The value of Λ_{costA} (i.e., actual after crash cost) was estimated by summing the individual PIRE costs for each crash in the after-period over all treated intersections in the jurisdiction. The value of VAR$[\Lambda_{costA}]$ was estimated by summing the variance for each individual cost of the crashes of interest in the after-period.

Π_{costA}, (i.e., the expected after cost without treatment) was estimated for a KABCO subgroup by first estimating an expected cost for each site as the product of Πi (i.e., the expected number of crashes in the KABCO subgroup) and the PIRE unit economic cost for the crash type, KABCO subgroup, and speed limit category. These were then summed over all treatment sites and KABCO subgroups to get Π_{costA}.

VAR$\{\Pi_{costA}\}$ for each site and subgroup was taken as product of Πi and the PIRE unit variance for the crash type, KABCO subgroup, and speed limit category. These variances were then summed over all sites and KABCO subgroups. This is an approximation that likely underestimates the variance, considering there is variance in the EB estimates of the expected number of crashes without treatment; however, the PIRE unit cost variances are also approximations because they do not include all components (e.g., variance in medical costs by diagnosis). Fortunately, the point estimates of the economic effects, which are of primary interest in this analysis, are quite insensitive to VAR$\{\Pi_{costA}\}$.

As noted, the theory so far applies for a given crash type of the three comprising all crashes. To obtain estimates of economic effect for all crash types combined, Λ_{costA}, VAR$\{\Lambda_{costA}\}$, Π_{CostA}, and VAR$\{\Pi c_{ostA}\}$ are first determined for each crash type as outlined and then summed over the three crash types before applying equations7 to 10.

IX. DATA COLLECTION

Following the approval of the basic research design described earlier, the project team initiated data collection in each of the seven jurisdictions. The data items listed in table 14 were sought in each jurisdiction. Team members visited each site and collected available raw data. They then coded the data into computerized analysis files based on protocols and data formats developed earlier.

Crash Data

Crash data for 3 or more years before RLC installation was sought for each treatment, reference, and control intersection in each city. Preliminary telephone interviews with each of the seven jurisdictions indicated that historic crash data were available in most cases; however, discussions held during the site visits indicated that while there were some computerized historical crash data available in some cities, most did not retain computerized (or raw) data for the needed before- treatment periods. The only exception was Charlotte, NC. There, staff could provide 1997 through March 2002 data in Microsoft®Access files from their current computer system. Earlier 1994–1996 data were not stored in the current system. Working with Charlotte staff, a computer analyst who had worked with the older files was located, and she was able to retrieve the old files and covert them to Access files for project use.

Neither the three jurisdictions in Maryland nor the three in California could provide adequate historic crash data because they either did not store multiple years of older data or were restricted as to what could be released. The California State Highway Patrol (CHP) is the repository of all police crash reports for the entire State. The project team requested and received computerized data from the CHP on all crashes for the three California jurisdictions for the years 1992–2002. Variables extracted for use in this analysis are described in the "Results" section.

The same data deficiency was found in the three Maryland jurisdictions. As noted earlier, the team had learned that the Maryland CODES team retained history crash data for all Maryland crashes. Project staff requested and received a computerized file of all crashes occurring in Howard and Montgomery Counties and Baltimore City for the period 1994–2001. Later, the additional data for 2002 were requested and received.

Intersection Inventory and Volume Data

As indicated, all available raw intersection geometric, signalization, and traffic flow data listed in table 10 were collected from the individual jurisdictions and coded to analysis files by the project team. During the early part of the data collection phase, it became apparent that the effort required and the cost of the data collection was going to greatly exceed the original estimate.

This collection and coding required both an onsite visit to each jurisdiction where the project team met with transportation staff and collected the raw data and subsequent extraction and coding of the data into a computerized analysis file. A small part of the increased effort resulted from the addition of a limited number of required variables at the end of the experimental plan effort; however, most of the increase resulted because in most cities, little or none of the data except for the crash data are computerized. Thus, in addition to finding and copying (and later coding) paper files, the team often needed to manually extract data from Computer Aided Detector Design (CADD) files, intersection drawings, and aerial photographs held in different offices.

As the result of internal research team discussions and conversations with the FHWA task order manager, steps were taken to both minimize data collection costs and to examine alternative data collection levels. It was decided that the decision concerning final data collection levels for treatment, signalized reference, and unsignalized control sites would be based on the development of safety performance functions using the data from two cities where full coding was done. The SPFs would be developed with various combinations of variables to determine the minimum set of data items required.

The project team developed these SPFs based on data from El Cajon, CA, and Howard County, MD, analyzed the data, and developed recommendations for FHWA review. (A copy of the internal memo describing this effort is available from the project team or FHWA.)

The findings indicated that the project would require what amounted to full coding of all traffic signal data, coding for all available traffic volume data (which is limited in some jurisdictions), and almost full coding of the intersection geometry data. More specifically, FHWA agreed with the project team that the final analysis files would include the following information:

- All pertinent crash data linked to the appropriate treatment, signalized reference, and unsignalized reference intersections.
- For RLC (treated) intersections, full coding of all non-crash data including coding of signal data (signal timing and changes over the project period), signal data related to left- turn protection, and data related to actuated versus fixed cycle length. Where available, data on light emitting diode (LED) and backplate presence were coded. All data related to the intersection geometry shown in table 14 were coded, including lane and median measurements and speed limits where available. All available traffic volumes were coded and converted to AADTs for each approach. Information on the location of RLC warning signs was also coded.
- For signalized reference intersections, modified coding of all non-crash data. This included all data noted in the paragraph previous for LED and backplate data, and lane and median widths. In addition, the signal timing data to be coded included only indicator variables for left-turn protection and whether a signal is actuated or always works under a fixed cycle length.
- For unsignalized control intersections, modified coding of all non-crash data. This includes volume data as noted in the second paragraph. The geometry data includes number of approaches, number of approaching lanes on each approach (without turn or through designation), and which approaches are stop-sign controlled.

It is noted that the sources of different intersection inventory variables differed from jurisdiction to jurisdiction. In most cases, the basic data source was a paper file; however, in some cases, one or more of the jurisdictions did not have paper files, and alternative sources were found and used.

For example, in Baltimore, MD, electronic CADD drawings were not available for any intersections, and hard-copy drawings were only available for a limited number of intersections (i.e., less than 20 percent). There, aerial photographs were located in the city's geographic information system (GIS) office, and project staff coded intersection geometrics from those photographs to the extent possible. Appendix A provides information on the basic source of data for each of the key variables in each jurisdiction.

As indicated, annual traffic volume data for each approach at all intersections is important in the development of the SPFs and the subsequent analyses. As was the case with other inventory data, the full array of needed volume data did not exist in data files in the jurisdiction. Unlike State systems where traffic volumes (i.e., AADT) for each section of roadway are generally updated each year through a series of counts and estimating procedures, only limited volume data were found in city and county files. While there were a limited number of midblock traffic counts at set locations in some jurisdictions, most traffic data were collected in the form of intersection turning-movement counts, and done as-needed. Thus, the available counts were often either multiple-hour turning movements (e.g., 4-hour or 11-hour counts), or such counts had been converted by the jurisdiction to average weekday daily traffic (AWDT) counts.

Thus, it became clear early in the data collection effort that estimates of annual approach volumes would need to be developed by the project team from the available count information. The types of available data, and thus the gaps in annual approach counts, varied from jurisdiction to jurisdiction. In all cases, the requirement was that there be at least one count for all approaches in either the before- or after-period, with the goal being to have at least one full count in each period.

Following is a list of the methods used to develop AADTs and fill in gaps:

- In limited cases, missing counts for city or county intersections were obtained from the State Department of Transportation (if the road was a State-system highway), or from local metropolitan planning organizations (MPOs).
- When hourly turning movement counts or AWDTs were available, they were converted into AADTs using factors either supplied by the jurisdiction (often on the count record) or factors developed by the project team based on information provided by the jurisdiction or the State Department of Transportation for that jurisdiction.
- When only one traffic count was available for the before- or after-period, that AADT was assigned to all years in that before-or after-period. (We chose not to use estimated growth factors between years because none were provided by the jurisdictions, and the project team felt that the presence of multiple intersections with counts in different years would allow the development of sufficient SPFs for the analysis.)
- When counts in differing years in either the before- or after-period existed for a given approach, the approach counts were averaged and that average was used for all years in either the before- or after-period.
- When only one count was available for an intersection, the AADT developed from it was assigned to all years in the before-and after-

periods. This occurred in less than 10 percent of the total intersections in the study.

- When an approach count existed for one approach but not the opposing one (usually for the minor roads), the opposing approach was made equal to the existing count.
- When a count of entering vehicles for the full intersection was found, but not individual approach counts, data from other years when approach-specific data were available at the same intersection were used to calculate a percentage at each approach, and the entering-vehicle counts were distributed using those percentages. (The project team analyzed one limited set of intersections where both entering vehicles and approach-specific counts were present for different years. The individual approach counts were predicted based on proportions from another year, and then compared to actual counts. There was fairly good agreement, with approach count errors usually between 5 percent and 20 percent. The larger percentage errors were usually associated with low counts, and thus did not represent large differences in frequencies. Given this level of agreement, the fact that most large errors seem to be for small counts, and the fact that the only alternative was to delete these intersections without suitable replacements, the project team felt this method was suitable.)

The AADT data ultimately developed and used in these analyses cannot be considered as accurate as what might be found in some State files. All available sources for additional data were explored; however, the project team considered the AADT data to be sufficient for use.

X. Data and SPF Preparation

Before the actual data analyses, preliminary efforts involving file merging and data quality checks were completed.

Crash Data Linkage to Intersections

In most State DOT crash and inventory files, as is the case with the State data in FHWA's Highway Safety Information System (HSIS), crashes can be computer-linked to inventory and traffic volume data for roadway segments and intersections using location reference variables such as route or milepost on each record. This was possible in Howard and Montgomery Counties, where project staff were provided an electronic file of the milepost book used by police officers in the field. The treatment and reference intersections were identified and matched to crashes based on this milepost information.

This was not the case for the intersections analyzed in other jurisdictions. In Charlotte, NC, the 1997 and later data included intersection control numbers for all intersections and crashes that could be used for file linkage. There were also control numbers in the pre-1997 data, but they differed from the later data. The Charlotte staff provided us with conversions between the new and old systems.

For all data for the three California jurisdictions and Baltimore, MD, no such location system existed, and the crashes had to be manually linked to pertinent intersections based on the names of the crossing streets. The crashes were sorted by street names and an analyst matched the crash-report streets with the street names from the treated and comparison intersection file. All combinations of crash-report street names were checked to pick up possible misspellings by the investigating officer or coder.

The project team was able to conduct a limited verification of both the completeness of the State CODES data files and the manual linkage procedures using El Cajon, CA, data. The local traffic engineer sent the project team crash summary reports for one treatment, one signalized reference, and one unsignalized control intersection. These summary reports contain a listing of all cases that have been coded to an intersection by city staff, using their own coding scheme. The comparison of these crashes to those identified and linked by the project team indicated that use of the State data resulted in minor differences with the local crash summaries. It was thus concluded that the State data is of sufficient quality.

Defining Red-Light-Running Crashes

As indicated earlier, the basic analyses were to be focused on target crashes, those red-light- running crashes that could be affected by the RLC treatment. The analysis would also examine other intersection crashes to confirm that unanticipated effects were not present. Because there is no "red-light-running" crash category on most police crash forms, these target crashes must be defined based on variables on the form. Definitions could range from only crashes in which a citation for a traffic signal violation was given to all right-angle crashes and rear end crashes at or near the intersection. One could choose to include rear end crashes that were noted by the officer as "intersection-related" (where this variable was present), or to include all rear end crashes approaching the intersection within a specified distance of the intersection. Depending on the distance (X) chosen, the assumption would be that the RLC would affect behavior of the lead vehicle or vehicles, which could result in rear end crashes X distance back in the approaching queue of vehicles. One could also choose to include left-turn opposite-approach crashes because some of these would be red-light-running crashes if a protected signal phase existed. To further complicate matters, the different jurisdictions use slightly different definitions of right-angle crashes on the report form.

Based on definitions used in previous studies, available data variables in the current files, and project team discussions, the following general decisions were made by the project team:

- In general, "RLC-related crashes" would include crashes in the intersection itself where one vehicle is "running the light," plus intersection-related rear end crashes that could be affected by RLC systems, including those rear end crashes occurring in the approach queue. Clearly, neither of these two types of crashes is explicitly defined in crash data. Thus, the following definitions were used.
- "Red-light-running" crashes at the intersection proper were defined as "angle," "broadside," or "right- or left-turning" crashes involving two vehicles, with the vehicles entering the intersection from perpendicular approaches. "Perpendicular approaches" was defined using the compass directions of travel for each involved vehicle, a variable that was present in the data for all seven jurisdictions. In most jurisdictions, all crashes meeting these "crossing" criteria and occurring at or within 6.096 m (20 ft) of the intersection were captured. (A second definition of these RLR crashes includes crashes involving a left-turning and a through vehicle from opposite approaches on the same roadway. This would capture those vehicles running the red signal either during or before or after a protected signal phase.)
- Rear end crashes used in the analyses were those defined as "rear end" by the crash type and occurring on any approach within 45.72 m (150 ft) of the intersection. Total intersection-related crashes were also analyzed. The definitions for each jurisdiction appear in table 15.

As could be expected, available crash variables and codes differed between cities, making it impossible to have totally consistent definitions across all seven jurisdictions. For example, only the three Maryland jurisdictions had an "intersection-related" code that can be used to further screen rear end crashes occurring within 45.72 m (150 ft) of the intersection. Thus, all rear end crashes within 45.72 m (150 ft) were used in Charlotte, NC, and all three California databases.

In addition, we encountered significant problems with the distance-from-intersection data in Baltimore, MD. Approximately 10 percent to 15 percent of the data appear to have questionable distances such as distances of 0.03 m (0.1 ft) and 0.30 m (1.0 ft) from the intersection. The project team attempted to verify these distances by obtaining hardcopies, but found that the accident case numbers in the computerized CODES data were not the same as the Baltimore Police Department case numbers, and only Baltimore has hardcopies of the reports. Thus, in the Baltimore analyses, two sets of data were used, a first set containing only rear end crashes within 45.72 m (150 ft) where the distance data were believed to be accurate, and a fuller set that also included crashes coded as within 45.72 m (150 ft), where the distance measurements were questionable. The analyses of these two sets of data revealed no significant differences; therefore the full set including the questionable distances was used for the final analysis.

The final set of criteria for each RLC-related crash type for each jurisdiction is listed in table 11.

Table 11. Definitions of crash types used in the analyses for each jurisdiction

El Cajon, San Diego, San Francisco, CA
Intersection-related —All crashes at or within 6.096 m (20 ft) of intersection; rear end crashes within 45.72 m (50 ft).
Right-angle 1 (RA1)—Broadside, head-on, or sideswipe where vehicles approach intersection from perpendicular directions. (California does not have "left-turn" or "right-turn" as a crash type. Because there could be crashes from perpendicular directions where one of the vehicles is turning, it is assumed that all turning crashes are coded as either broadside, head-on, or sideswipe.)

Table 11. (Continued)

Right-angle 2 (RA2)—Crashes in RA1 plus opposite direction left-turn. This may not be as precise a definition as RA1 because it could include non-RLR crashes in which the oncoming vehicle and the turning vehicle both had a green signal. That is, these are not restricted to locations with protected left-turn phases only. However, opposite direction left-turn crashes do include RLR crashes in which a vehicle turning left at the end of a green phase (referred to as a "sneaker" in traffic engineering terminology) is broadsided by a vehicle from the opposing direction that is technically running a red light. *Rear end*—All rear end crashes within 45.72 m (150 ft) of intersection.
Charlotte, NC *Intersection related*—All crashes at or within 6.096 m (20 ft) of intersection; rear end crashes within 45.72 m (150 ft) of intersection. *Right-angle 1 (RA1)*—Angle, head-on, sideswipe, left-turn different roadways, right-turn different roadways where vehicles approach intersection from perpendicular directions. *Right-angle 2 (RA2)*—Those in RA1 plus opposite direction left-turn. *Rear end* —All rear end crashes within 45.72 m (150 ft) of intersection.
Howard County, Montgomery County, MD *Intersection-related*—All crashes within 48.158 m (158 ft) and identified as "intersection" or "intersection- related." *Right-angle 1 (RA1)*—Vehicles approach intersection from perpendicular directions, in any category of head-on, head-on left-turn, opposite direction sideswipe, straight movement angle, angle meets right- turn, angle meets left-turn, or angle meets left head-on. *Right-angle 2 (RA2)*—Those in RA1 plus opposite direction left-turn. *Rear end*—All rear end crashes within 48.158 m (158 ft) and identified as "intersection" or "intersection- related."

Baltimore, MD

Intersection–related—All crashes within 48.158 m (158 ft) and identified as "intersection" or "intersection- related."

Right-angle 1 (RA1)—Vehicles approach intersection from perpendicular directions, in any category of head-on, head-on left-turn, opposite direction sideswipe, straight movement angle, angle meets right- turn, angle meets left-turn, or angle meets left head-on.

Right-angle 2 (RA2)—Those in RA1 plus opposite direction left-turn.

Rear end—All rear end crashes within 48.158 m (158 ft) and identified as "intersection" or "intersection- related."

Development of Safety Performance Functions

As indicated earlier, the study required the development of safety performance functions (SPFs) for signalized and stop-controlled intersections. A reference group of untreated signalized intersections was used to develop SPFs to account for traffic volume changes and regression to the mean using the empirical Bayes procedure. The unsignalized intersection SPFs were used to account in that procedure for time trends in crash counts unrelated to the RLC installation.

Therefore, it was necessary to first ensure that the comparison group used to calibrate the SPFs was suitable for this purpose, that is, that it had similar crash trends to the treatment group over the years before RLC installation. To this end a comparability test as outlined in Hauer was performed [4]. This test confirmed the suitability of the comparison group.

To build the strongest possible SPFs, reference group data (i.e., data from the untreated signalized intersections) were combined for sets of jurisdictions, considering proximity and similarity in crash reporting practices. To this end, the three California cities of El Cajon, San Diego, and San Francisco were combined.

Not only are these three cities in proximity, but they also do not have full reporting of PDO crashes, and the crash data all came from the State database maintained by the CHP.

Howard and Montgomery Counties, MD, reference group data were combined because of their proximity and similarity in reporting practices.

Baltimore, MD, and Charlotte, NC, were combined because of their high reporting of non-injury crashes. In each case where jurisdictions were combined, a jurisdiction-specific multiplier was calibrated and applied to account for any remaining differences in crash reporting.

Development of the SPFs involved determining which explanatory variables should be used, whether and how variables should be grouped, and how variables should enter into the model, in other words, the best model form. Generalized linear modeling was used to estimate model coefficients using the software package GENSTAT and assuming a negative binomial error distribution, all consistent with the common recent research practice in developing these models [29].

In specifying a negative binomial error structure, the dispersion parameter, *k*, which relates the mean and variance of the regression estimate, is iteratively estimated from the model and the data. The value of *k* is such that the smaller its value, the better a model is for a given set of data.

For specific crash types at signalized intersections, a multiplier is applied to the model that is equal to the proportion of total crashes that each crash type makes up.

A value of *k* was calculated for each crash type using a maximum likelihood process, as explained earlier. Similarly, although data for groups of jurisdictions were combined for SPF calibration, separate multipliers and *k* values were calculated for each jurisdiction.

The inclusion of variables such as number of lanes rarely significantly affected the fit. This is not surprising because, as previous research has shown, much of the variation in crash experience is explained by the volume of traffic entering an intersection.

The results of the SPF calibration for the signalized reference group are presented in table 12. The model forms used are tried and tested and, because of the limited datasets available, options on model forms and variables to include were so limited that a trial and error modeling approach, using published models as a guide, was realistic. In addition, fine tuning the model is not critical in EB analysis, especially because by weighting the observed count, one is accounting for omitted variables that may affect crash frequency.

Table 12. Safety performance functions for the signalized intersections reference group

3-legged							
	El Cajon	San Diego	San Francisco	Howard Co.	Montgomery Co.	Baltimore	Charlotte
Model form crashes/year	$\alpha(F1+F2)^b$			$\alpha(F1+F2)^b exp(minllane*e)$		$\alpha(F1)^c(F2)^d exp(majllanc*f)$	
Ln(α) (s.e.)	-5.240 (2.21)	-5.651 (2.22)	-5.240 (2.21)	-6.970 (1.800)	-6.970 (1.800)	-3.100 (1.240)	-3.100 (1.240)
B (s.e.)	0.580 (0.218)	0.580 (0.218)	0.580 (0.218)	0.709 (0.183)	0.709 (0.183)	–	–
C (s.e.)	–	–	–	–	–	0.374 (0.119)	0.374 (0.119)
D (s.e.)	–	–	–	–	–	0.136 (0.080)	0.136 (0.080)
E (s.e.)	–	–	–	0.964 (0.297)	0.964 (0.297)	–	–
F (s.e.)	–	–	–	–	–	0.264 (0.075)	0.264 (0.075)
Total α, k	1.00, 0.18	1.00, 0.28	1.00, 0.28	1.00, 0.30	1.00, 0.30	1.00, 0.56	1.00, 0.28
Injury α, k	0.28, 0.13	0.31, 0.26	0.26, 0.26	0.12, 0.30	0.24, 0.21	0.15, 0.91	0.07, 0.24
Right-angle α, k	0.40, 0.67	0.35, 0.91	0.55, 0.91	0.35, 0.37	0.28, 0.14	0.44, 1.0	0.25, 0.45
Rear end α, k	0.41, 0.18	0.43, 0.25	0.22, 0.25	0.39, 0.63	0.44, 0.03	0.18, 1.1	0.61, 0.45
4-legged							
	El Cajon	San Diego	San Francisco	Howard County	Montgomery County	Baltimore	Charlotte
Model form crashes/yr	$\alpha(F1+F2)^b exp(minrlane*e)$			$\alpha(F1)^c(F2)^d$		$\alpha(F1)^c(F2)^d exp(majllane*f)$	
Ln(α) (s.e.)	-3.950 (2.010)	-4.624 (2.021)	-4.477 (2.021)	-8.370 (1.090)	-8.370 (1.090)	-3.100 (1.240)	-3.100 (1.240)
B (s.e.)	0.530 (0.197)	0.530 (0.197)	0.530 (0.197)				

Table 12. (Continued)

3-legged							
	El Cajon	San Diego	San Francisco	Howard Co.	Montgomery Co.	Baltimore	Charlotte
Model form crashes/year	$\alpha(F1+F2)_b$			$\alpha(F1+F2)_b \exp(minllane*e)$		$\alpha(F1)_c(F2)_d \exp(majllane*f)$	
C (s.e.)	–	–	–	0.703 (0.103)	0.703 (0.103)	0.374 (0.119)	0.374 (0.119)
D (s.e.)	–	–	–	0.335 (0.075)	0.335 (0.075)	0.136 (0.080)	0.136 (0.080)
E (s.e.)	-0.279 (0.129)	-0.279 (0.129)	-0.279 (0.129)	–	–	–	–
F (s.e.)	–	–	–	–	–	0.264 (0.075)	0.264 (0.075)
Total α, k	1.00, 0.19	1.00, 0.24	1.00, 0.24	1.00, 0.20	1.00, 0.20	1.00, 0.56	1.00, 0.28
Injury α, k	0.26, 0.14	0.29, 0.10	0.26, 0.10	0.16, 0.20	0.25, 0.25	0.15, 0.91	0.07, 0.24
Right-angle α, k	0.48, 0.34	0.42, 0.38	0.55, 0.38	0.38, 0.36	0.48, 0.45	0.44, 1.0	0.25, 0.45
Rear end α, k	0.32, 0.33	0.39, 0.48	0.22, 0.48	0.40, 0.45	0.32, 0.24	0.18, 0.9	0.61, 0.45

Table Legend:

F1 = entering AADT on major road, *F2* = entering AADT on minor road; *minllane* = number of left-turn lanes on the minor road; *majllane* = number of left-turn lanes on the major road; *minrlane* = number of right-turn lanes on the minor road; *(s.e.)* = standard error of the estimate; *k* is a calibrated parameter relating the mean and variance used in the empirical Bayes estimation procedure.

XI. RESULTS

Results of this study were obtained separately for the composite effects at the camera sites and reference sites analyzed for spillover effects. As best as could be determined, RLC installation was the only change that occurred at the sites when the cameras were installed. In a few cases, the intersection was substantially changed in either the before- or after-period. In such cases, the data after a change in the after-period and before a change in the before-period were excluded from the analysis; therefore, the reported effects most likely result from RLC installation.

Composite Effects at Camera sites

Because the intent of the research was to conduct a multijurisdictional study representing different locations across the United States, the aggregate effects over all RLC sites in all jurisdictions was of primary interest. Table 13 shows the combined results for the seven jurisdictions. As seen, there is a significant decrease in right-angle crashes but a significant increase in rear end crashes. Note that "definite injury" crashes here are defined as K, A, and B crashes; they do not contain the "possible injury" crashes captured by KABCO-level "C." Table 14 indicates that, while the magnitude of the effects changes across jurisdictions, the direction of the effects is remarkably consistent.

Table 13. Combined results for the seven jurisdictions

	Right-angle		**Rear end**	
	Total crashes	**(Definite) injury**	**Total crashes**	**(Definite) injury**
EB estimate of crashes expected in the after-period without RLC	1,542	351	2,521	131
Count of crashes observed in the After-period	1,163	296	2,896	163
Estimate of percentage change (standard error)	- 24.6 (2.9)	- 15.7 (5.9)	14.9 (3.0)	24.0 (11.6)
Estimate of the change in crash frequency	- 379	- 55	375	32

Note: A negative sign indicates a decrease in crashes.

Table 14. Results for individual jurisdictions

Jurisdiction number (in random order)[a]	Right-angle crashes	Rear end crashes
	Percentage change (standard error)	Percentage change (standard error)
1	- 40.0 (5.4)	21.3 (17.1)
2	+ 0.8 (9.0)	8.5 (9.8)
3	- 14.3 (12.5)	15.1 (14.1)
4	- 24.7 (8.7)	19.7 (11.7)
5	- 34.3 (7.6)	38.1 (14.5)
6	- 26.1 (4.7)	12.7 (3.4)
7	- 24.4 (11.2)	7.0 (18.5)

[a] Jurisdictions are not identified because of an agreement with them, and it is irrelevant to the findings.

Note: A negative sign indicates a decrease in crashes.

Spillover Effects

To investigate possible spillover effects of RLC programs, a separate analysis was performed using the untreated signalized intersection reference sites. For this analysis, the before-and after- periods for these sites in each jurisdiction were demarcated by the year of the first RLC installation at the treatment sites. (Because, by definition, specific treatment dates do not exist for each untreated reference site, this decision was based on the assumption that the public may have perceived that cameras were at noncamera locations from the time of the initial publicity campaign.) Table 15 shows the composite results of this analysis combining data from all of the jurisdictions. As seen, there are indications of a modest spillover effect on right-angle crashes. That this is not mirrored by the increase in rear end crashes that was seen in the treatment group would detract somewhat from the credibility of this result as evidence of a general deterrence effect.

Discussion of Crash Effects

This statistically defendable study found effects that were consistent in direction with those found in many previous studies, although the benefits were somewhat lower than those reported in many sources. This indicates that regression to the mean might have been at play in many of those studies, and it emphasizes the need for controlling those effects in an evaluation of red- light-camera programs, and studies of road safety countermeasures in general.

The opposite direction effects for rear end and right-angle crashes deserves attention from two perspectives. First, the extent to which the increase in rear end crashes negates the benefits for right-angle crashes is unclear at this point. An examination of the changes in crash numbers is insufficient to provide clarity on this issue because of differences in severity levels between right-angle and rear end crashes and in the changes in these crashes following RLC installation. The economic analysis, discussed in the next section, examines the economic costs of the changes based on an aggregation of rear end and right-angle crash costs for various severity levels.

Table 15. Before-and-after results for total crashes at spillover intersections

	Right-angle crashes	Rear end crashes
EB estimate of crashes expected in the after-period without RLC	3,430	3,802
Count of crashes observed in the After-period	3,140	3,873
Estimate of percentage change (standard error)	- 8.5 (2.2)	1.8 (2.3)

Note: A negative sign indicates a decrease in crashes.

The second perspective of the opposing effects for the two crash types is the implication that RLC systems would be most beneficial at intersections where there are relatively few rear end crashes and many right-angle ones. To provide better guidance on this issue requires an examination of the net economic effect on intersections grouped by the numbers of each crash type. That examination, which seeks generally to isolate all of the factors that would favor (or discourage) the installation of RLC systems, uses the net economic benefit as the outcome variable, also discussed in the next section.

The indications of a spillover effect point to a need for a more definitive study of this issue. That more confidence could not be placed in this aspect of the analysis reflects that this is an observational retrospective study of RLC installations that took place over many years, and in locations where other programs and treatments may have affected crash frequencies at the spillover study sites. A prospective study with an explicit purpose of addressing this issue appears to be required.

The Economic Analysis of RLC-Related Severity and Frequency Changes

Development of Unit Crash Cost Estimates

This study needed economic cost per crash for the categories of interest. Those categories included "right-angle," "rear end", and "other" at urban and rural signalized intersections. The crash cost needed to be keyed to police crash severity (KABCO) found in the files available for use. In addition, because of limited sample sizes for fatal and severe (A) injury crashes in the after-period for some study intersections, crash costs were needed for combined categories such as K+A severity.

The Pacific Institute for Research and Evaluation (PIRE) developed the cost estimates used in this RLC analysis as part of a larger effort of producing cost estimates for other crash types. Details of the development of the unit crash-cost estimates can be found in a recent paper and in an internal report available from FHWA.(5,6). In summary, by merging previously developed costs per victim keyed on the AIS injury severity scale into U.S. traffic crash data files that scored injuries in both AIS and KABCO scales, PIRE economists could produce estimates for both economic (human capital) costs and comprehensive costs per crash. The comprehensive cost estimates include both economic costs and costs associated with losses in the quality of life.

In addition, the analysis produced an estimate of the standard deviation and the 95-percent confidence intervals for each average cost. Following is a list of databases PIRE used:

- The 1999–2001 Crashworthiness Data System data provided the basic data source. This database includes both AIS and KABCO injury scaling for passenger vehicle occupants in towaway (but not nontowaway) crashes [30]. Note that nontowaway crashes would be predominately noninjury, "O" crashes.
- The 1982–1986 National Accident Sampling System data were used to fill in the nontowaway part of the distribution [31]. While the data were not recent, the information provided the most recent medical description available on injuries to other non-CDS crash victims.

- The 1999–2001 GES data were then used to weight the NASS data so that they represent the annual estimated GES injury victim counts in non-CDS crashes [32]. Applying this information controlled for crash type (as defined by geometry), police-reported injury severity, speed limit (≤ 72.42 km/h [45 mi/h] and > 80.47 km/h [50 mi/h]), and restraint use. Weighting the NASS data to GES restraint-use levels updates the NASS injury profile to a profile reflecting contemporary belt use levels. Sample size considerations drove the decision to pool 3 years of data. At the completion of the weighting process, a hybrid CDS/NASS file had been developed that included weights that summed to the estimated current annual incidence by police-reported injury severity and other relevant factors.

To meet the needs of this project and future FHWA projects, both comprehensive and human capital cost estimates were developed for six KABCO groupings within 22 selected crash types and two speed limit categories (≤ 72.42 km/h [45 mi/h] and > 80.47 km/h [50 mi/h]). As indicated earlier, the KABCO groupings ranged from detailed estimates for each level of crash severity within each crash type to combined levels of KABCO without regard to crash type. All estimates were stated in Year 2001 dollar costs.

Because this RLC analysis involved placing a value on fatal crashes, comprehensive cost estimates (which include quality-of-life losses) were used as recommended in Council, et al. [3] This specific analysis was focused on right-angle and rear end crashes at signalized intersections in urban areas. Speed limits of ≤ 72.42 km/h (45 mi/h) and > 80.47 km/h (50 mi/h) were used as surrogates for urban and rural here because it was not possible to define an urban/rural variable in the databases PIRE used. Even though a limited number of the study locations had speed limits of 80.47 km/h (50 mi/h), urban unit costs were assigned to all crashes in the analysis. The effect of this approximation probably is small because only 10 of the 132 sites had speed limits of 80.47 km/h (50 mi/h) or more, and all were located in urban areas. Because the initially developed cost estimates for B- and C-level rear end crashes indicates some anomalies in the order (e.g., C-level cost were higher, probably because on-scene police estimates of "minor injury" often ultimately include expensive whiplash injuries), the B- and C-level costs were combined by PIRE into one cost. In initial economic analysis, an attempt was made to use three cost categories within each of the pertinent crash types, K+A, B+C, and no-injury. (It is not feasible to analyze fatal injuries separately in a study such as this because there were limited fatal crashes in any time period. The cost of one fatal crash in any cell could significantly bias the results.) However, because of the low sample sizes of fatal and serious (A-level) crashes in the after-period for some intersections, and the need to use the same cost categories across all intersections in all seven jurisdictions, two crash cost levels were ultimately used in all analyses, injury (K+A+B+C) and non-injury (O). The original estimate developed by PIRE and the combined weighted cost-per-crash estimates used for each crash type are shown in table 16. Also shown in table 16 are the standard deviations for the 2 severity categories used in the analyses.

Table 16. Original comprehensive crash cost estimates for urban signalized intersections by severity level and combined weighted estimates used in the economic effects analysis

Crash severity level	Right-angle crash cost	Rear end crash cost
K	$4,090,042	$3,781,989
A	120,810	84,820
B	103,468	27,043
C	34,690	49,746
O (standard deviation)	8,673 (1,285)	11,463 (3,338)
K+A+B+C "injury crash" (standard deviation)	$64,468 (11,919)	$53,659 (9,276)

Empirical Bayes Estimates of the Economic Effects

Table 17 gives the results for the economic effects including and excluding PDO crashes, estimated from equations 7 to 10 and the associated procedures shown earlier. The latter estimates are included because several jurisdictions considerably underreport PDO collisions. The columns labeled "All crashes" include non-right-angle, nonrear end "other" crashes for which reliable unit costs could not be developed by PIRE because of small sample sizes. It was decided that the same costs as for angle crashes would be used for the "other" category. For completeness, the small changes in these other crashes needed to be accounted for in reporting effects on all crashes, even though the changes may be random and have nothing to do with RLC installation.

Table 17. Economic effects* including and excluding PDOs

	All severities combined			PDOs excluded		
	Right- angle crashes	Rear end crashes	All crashes	Right- Angle crashes	Rear end crashes	All crashes
EB estimate of crash costs without RLC	$66,814,067	$69,347,624	$161,843,021	$61,687,367	$52,681,148	$134,407,104
Cost of crashes recorded after RLC (370 site years)	$48,319,090	$75,222,780	$147,470,550	$43,868,392	$53,944,539	$115,901,685
Percentage change in crash cost (s.e.) [negative is decrease]	- 27.7 (0.6)	8.5 (0.7)	- 8.9 (0.4)	- 28.9 (0.6)	2.4 (0.8)	- 13.8 (0.5)
Crash cost decrease (per site year)	–	–	$14,372,471 ($38,845)	–	–	$18,505,419 ($50,015)

*Using a combined unit cost for K+A+B+C

The results show a positive aggregate economic benefit of more than $14 million over approximately 370 site years, which translates into a crash reduction benefit of approximately $38,000 per site year. The implication from this result is that the lesser severities and generally lower unit costs for rear end injury crashes together ensure that the increase in rear end crash frequency does not negate the decrease in the right-angle crashes targeted by red-light-camera systems.

As noted earlier, sample size considerations forced the combination of all injury crashes into one category (K+A+B+C). Concern was raised that the distribution of crash severity within this combined category might have changed between the before-and-after periods for either or both crash types. That is, injury-related angle crashes could have become more or less severe between the two periods. If so, the use of one injury-crash cost for both periods would be questionable. Table 18 presents the distributions of both right-angle and rear end injury crashes in each period.

Table 18. Severity-level distributions for right-angle and rear end injury crashes in the before-and-after periods

Crash type	Total frequency	Percentage in each Injury category			
		K	A	B	C
Right-angle					
Before	1,854	0.5	7.7	30.8	61.1
After	634	0.8	8.5	37.4	53.3
Rear end					
Before	1,930	0.1	3.1	10.9	86.0
After	1,008	0.0	2.7	13.5	83.8

As can be seen, while there is no apparent shift in the severity distribution for rear end injury crashes from before to after, the right-angle crashes appear slightly more severe in the after- period (i.e., the percentages of K, A, and B are slightly higher). Further analysis revealed that this shift occurred in only two of the seven jurisdictions. Nevertheless, because the same crash costs essentially were used for before-and after-periods, this means that the cost of the after- period right-angle crashes may be slightly underestimated, even when all jurisdictions are combined. An attempt was made to estimate the potential effect of this shift on the economic savings, even though this could be done only by using anomalous data for individual KABCO categories that we argued against using earlier. With these data, it appears that if the shift were real, the overall cost savings reported in the last row of table 17 could be decreased by approximately $4 million; however, note that there would still be positive economic benefits, even if it is assumed that the unit cost shifts were real and correctly estimated.

Examination of the aggregate economic effect per after-period year for each site indicates substantial variation, much of which could attributed to randomness. It was reasonable to suspect that some of the differences may be to the result of factors that affect RLC effectiveness. The results of the examination of those factors are described next.

Factors Affecting RLC Effectiveness

Two types of disaggregate analyses were undertaken to identify factors associated with the greatest economic benefits or that might discourage the use of RLCs. The basic outcome measure used is the aggregate economic effects, that is, the combined economic effects on rear end, right-angle, and other crashes of various severities. The economic effect for each crash type and severity was derived from equation 5 as the difference between the expected cost of crashes in the after-period if no RLC were installed and the cost of crashes actually occurring at the treatment sites in the after-period.

The first analysis was a univariate exploration of the results of aggregate economic effects for each intersection, aiming to identify factors that might be associated with the variation in the effects at individual sites. In this, spreadsheets were used to sort the data and results for each site by various columns, and to group by ranges of a variable to explore the relationship between factors and the measured aggregate economic effect per after period site year for a group as a whole for all crash types combined. For example, sorting by ascending order of AADT, the spreadsheet was set up so that aggregate economic effect per after-period site year in a given row applies to all sites with AADT less than or equal to the value in that row; one can then look for trends in the outcome measure. Similarly, by sorting by publicity level, one can obtain the aggregate economic effect per after-period site year separately for the 85 sites with high publicity level and the 47 sites with medium publicity level.

Naturally, some of the conclusions from the univariate exploratory analysis could result from correlation among the various variables found to affect the RLC effect. This could mask the effects or indicate effects that are not real. The results of the exploratory analyses were used to guide a more formal analysis that used multivariate modeling to assess whether the conclusions from the univariate analysis might remain despite the obvious correlations among variables found in that analysis to be associated with economic effects.

In this more formal disaggregate analysis, data for all jurisdictions were combined to develop a model to estimate the value of aggregate economic effect per site year for an individual site using traffic volumes and other site characteristics, such as proportion of rear end or right-angle crashes and signalization features, and RLC implementation features such as publicity level as explanatory variables. The model was a linear one with a normal error distribution. It took the following form:

$$\Phi_{cost}\, per\ after\ period\ year = \alpha + b_1 x_1 + b_2 x_2 + b_3 x_3 + \ldots\ b_n x_n \tag{11}$$

where α is the calibrated intercept and b_1, b_2, … , b_n are the estimated effects on Φ_{cost} *per after- period year* of factors x_1, x_2, x_3 ….x_n.

Stepwise linear regression was performed with the SAS statistical analysis software package, using the estimates of the Φ_{cost} *per after-period year* as estimates of the dependent variable. It should be pointed out that the absence of a variable in the final model does not necessarily mean that the variable would not affect the safety effect of RLCs because an effect with low statistical significance could result from correlation with other variables, a lack of variation in the data, or a sample that is too small. In addition, it should be emphasized that the generally small size of the aggregate economic effect of RLCs was already strongly indicative of the reality that one is unlikely to detect with significance many factors that affect the safety effect of RLCs.

Data for all treated intersections in all seven jurisdictions were used in this analysis. However, the different jurisdictions had different crash reporting thresholds, which resulted in significantly different numbers and percentages of non-injury crashes across jurisdictions. Because this analysis required that the crash costs for all intersections (and thus all jurisdictions) be calculated on a common basis, non-injury crashes were omitted from this analysis. Because the analysis is aimed at identifying factors of interest, and because these factors can be identified as logically with injury crashes as with total crashes, this was felt to be proper procedure.

The exploratory univariate analysis led to the following general observations on the net economic effects:

- High publicity level (85 sites) is associated with a greater benefit than medium publicity level (47 sites).
- Fine plus demerit point penalty (90 sites) is associated with a greater benefit than a fine- only penalty (42 sites).
- Warning sign at intersections only (39 sites) is associated with a smaller benefit than warning sign at both intersections and city limits (73 sites).
- Benefits are greater at sites with one or more left-turn protected phases (105 sites) than at those with no protected phases (27 sites). This variable may well be a surrogate for the volume of left-turning traffic or opportunities for crashes involving a vehicle going straight through and one turning left at the end of a protected or permitted phase.
- There are indications that the aggregate economic benefit increases with total entering AADT, increasing proportion of total traffic being on the major road, and with an increasing ratio of right-angle to rear end crashes.
- There are indications that the aggregate economic benefit increases with shorter cycle lengths and shorter intergreen periods. These intuitive indications were derived despite the difficulty of defining these variables for a given intersection because of variation in them over the years and even over a single day. The maximum recorded values for these variables in the study period were used in the analysis in the absence of a more stable and pertinent measure of these factors.

Clearly some of these variables that indicate effects in the univariate analysis are correlated, and therefore they may show effects that are not real. For example, left-turn protection is likely related to traffic volume levels; high publicity levels may exist in jurisdictions with the highest traffic volumes. To mitigate this difficulty, the multivariate regression analysis was undertaken to see if the direction of the effect of a given variable remains the same if the effects of other variables are considered simultaneously. This additional analysis confirmed the direction of all of the effects observed except for the penalty variable and the one related to the presence of a left-turn protected phase.

In interpreting these results, consideration should be given to the following important points:

- Factors other that the ones previously identified were examined. These include traffic signal actuation, presence of turn restrictions, major road speed limit, and number of approach legs; for these, the inability to detect a clear-cut effect may have resulted from the small samples for one level of the factor (e.g., only 27 of the 132 sites had no protected left-turn phases).
- The intent of the multivariate regression analysis was to confirm the direction of the effect, not to establish effects with statistical significance or to assess the size of the effect. To undertake analyses for these purer purposes would have required a substantially larger database, much more precision in the estimate of economic effect at each site, and more accurate specification and measurement of the independent variables. For the purposes of this current investigation, it suffices that both the univariate and multivariate analyses are reasonably in accord with the perceptions that are commonly held by those involved in red-light-camera programs.
- Some of the variables may well be surrogates for others that more directly influence the aggregate economic effects. For example, the presence of left-turn protection probably is associated with the volume of left-turning traffic or, more directly, with opportunities for crashes involving a vehicle going straight through and one turning left at the end of a protected or permitted phase.
- The results do not provide numerical guidance for trading off the effects of various factors. The intent of identifying these factors is to assist RLC implementers in choosing sites for treatment installation and determining the type of signing and publicity that might enhance the results of the program. For site identification, the results indicate that an implementer should give the highest priority for RLC implementation to the sites with most or all of the positive binary factors present (e.g., left-turn protection) and with the highest levels of the favorable continuous variables (e.g., higher ratios of right-angle crashes to rear end crashes).
- Based on the combined univariate analyses and modeling, as well a logical consideration of the result of the crash effects analysis that rear end crashes increase and right-angle ones decrease following RLC implementation, it would appear that the most important determinant of site choice would be a high ratio of right- angle to rear end crashes. After site choices are made, signing at both intersection and city limits

and a high-level publicity campaign also appear to increase program benefits.

- To quantify the potential aggregate benefit for a contemplated RLC site, it is possible to use the SPFs and probable estimates of safety effect shown in table 17. The rudiments of the procedure are documented in two recent publications [23,33]. In that procedure, crash and traffic data at the intersection are used to obtain an empirical Bayes estimate of the expected number of crashes by impact type and severity without a red-light camera. The estimate of probable safety effect from table 13 is applied to the EB estimate to derive an estimate of the expected change in crashes per year by type with the RLC implemented. The cost per crash derived for this project can then be applied to the crash changes expected for each impact and severity type. The results can then be summed to obtain an estimate of the aggregate benefit per year for the contemplated installation.

XII. Discussion and Conclusion

Red-light running at signalized intersections is a significant problem in the United States; it results in more than 95,000 crashes and approximately 1,000 deaths per year. Red-light-camera systems aimed at reducing this problem have become a popular tool in local jurisdictions. Their use has not been without controversy, primarily related to the use of private firms to implement the program, and questions concerning changes in signal timing during program implementation. Part of the controversy has stemmed from the lack of sound research concerning the effects of RLCs on intersection crashes. Many studies of RLC effectiveness were conducted in jurisdictions outside the United States, and most of the U.S. and non-U.S. studies have experienced methodological problems, as was documented by the critical review of literature conducted in this effort. This current study was an attempt to overcome these methodological issues and to examine the crash-related effects in multiple U.S. jurisdictions to see if consistent results were found.

At the beginning of the study, the FHWA oversight panel defined a series of first-priority questions to be addressed. Following is a list of questions, followed by the study findings for each.

- *What effect do RLCs have on intersection safety (i.e., intersection crashes) at monitored intersections versus intersection safety throughout the jurisdiction?*

  The results of empirical Bayes crash-frequency analyses at the treated intersections indicate that RLCs have effects similar in direction but somewhat smaller in magnitude than those indicated in past studies. Right-angle crashes (the surrogate for "red-light- running" crashes) decrease significantly and rear end crashes increase. Table 17 shows the combined results from all seven jurisdictions, indicating a 24.6 percent reduction in total right-angle crashes and a 15.7 percent reduction in right-angle (definite) injury crashes. Total rear end crashes increased by 14.9 percent, and rear end (definite) injury crashes increased by 24 percent. While the results varied some across the seven jurisdictions, the direction and degree were remarkably consistent, particularly given the differences in crash-reporting practices between jurisdictions.

  The results were not as clear for effects at other signalized intersections in the same jurisdiction, known as the "spillover" effects. Here, a modest decrease in right-angle crashes was seen (table 15), but, because the results did not show the expected companion increase in rear end crashes, there is some question of whether the observed difference is to the result of "spillover," the difficulty in defining the before-and after- periods for these untreated signalized intersections (because the treatments in all jurisdictions stretched across multiple years), or to other changes at these untreated locations during the study period. It appears that a well-designed prospective study will be needed to more confidently establish any spillover effect from RLCs.

  Finally, because the decrease in right-angle crashes was coupled with an increase in rear end crashes, because there can be more rear end crashes at intersections, and because the severities of the two crash types differ, it was important to combine both frequency- and severity-related effects into one analysis to determine the overall effect of RLCs. This was estimated as the difference between the economic costs of crashes expected and observed in the after-period at the treated intersections, with the former cost based on the empirical Bayes expected crash frequency. Updated estimates of comprehensive and human capital costs per crash were developed for 22 crash types in this project, and those defined for right-angle, rear end, and other signalized intersection crashes were used in this analysis. The combined results from the seven jurisdictions indicated a positive aggregate economic benefit of approximately $39,000 per site per year when property- damage-only (PDO) crashes are included and $50,000 per site per year when PDO crashes are excluded (table 21). These results indicate that the increase in rear end crash costs (due to the increase in frequency, with a lower severity) do not negate the savings in right-angle crash costs.

- *What is the relationship of signal timing (i.e., length of the yellow interval, length of the all-red interval, and various combinations of the yellow interval and all-red interval) with safety at intersections with RLCs? Later discussion indicated that the key factors of interest are yellow interval, all-red interval, cycle length, and signal coordination. The basic issue related to yellow time is the nature of the yellow phase length—e.g., a standard length, length based on ITE recommendations related to approach speeds and other factors, or some variation of these. The basic question for all-red phases is whether or not there is one (i.e., presence or absence of all-red phase). Cycle length is needed both to provide some measure of the number of red phases (and thus the number of opportunities for red-light-running) in a given time period, but also because longer red phases might "induce" more red-light-running. With respect to signal coordination, the issue is whether the treated signal approach is part of a set of coordinated signals that lead to queuing of vehicles (but not any additional details of the level of coordination).*

The analysis efforts here were focused on identifying signal-related factors that could increase or decrease RLC effects. Because such factors can affect both angle and rear end crashes, the outcome was based on the aggregate economic effects per year for a given treatment site. Univariate analyses identified factors associated with higher economic benefit, and regression models were used to verify the direction of univariate effects. (It was not possible to use the models to define either relative size or statistical significance of the individual effects, because this would require a much larger database and the ability to more precisely link a given signal attribute with a given location. In the existing database, as would be the case for almost all signalized intersections, signal timing changes both across years and even within a given day.) In general, it appears that the aggregate economic benefit increases with total entering AADT, an increasing ratio of right-angle crashes to rear end crashes, an increasing proportion of total traffic being on the major road, shorter cycle lengths, and shorter intergreen periods, and is greater for locations with one or more protected left-turn phases as opposed to intersections without such protection. Other factors such as traffic signal actuation, signal coordination, presence of turn restrictions, major road speed limit, and number of approach legs were also investigated; for these, the inability to detect a clear-cut effect may have been caused by the small samples for one level of the factor.

It is again noted that determining the relative importance of each of these factors was not possible, partly because of the noted modeling issues, but also because many of the variables are highly correlated with each other. It is further noted that these findings, even though not as detailed as might be desired, do provide guidance for implementers who want to maximize the potential benefit of RLC programs through good site choice. Given a set of potential RLC locations, an implementer should give the highest priority for RLC implementation to the sites with most or all of the positive binary factors present (e.g., left-turn protection) and with the highest levels of the favorable continuous variables (e.g. higher ratios of right-angle crashes to rear end crashes). Based on the combined univariate analyses and modeling, as well a logical consideration of the result of the crash effects analysis that rear end crashes increase and right-angle ones decrease following RLC implementation, it would appear that the most important determinant of site choice would be a high ratio of right-angle crashes to rear end crashes.

- *Are there certain improvements (e.g., signal timing, signage, geometric changes, etc.) done in conjunction with RLC installation that make the automated enforcement program more or less effective? Later discussion of the signage issue indicated that the key question has to do with presence or absence of "warning" signage, whether the sign is located at the intersection or away from the intersection (e.g., at the edge of town or at the beginning of a corridor), and whether informational signs providing data on the number of violations issued are used (because such signs have been shown to increase the effect of seatbelt enforcement programs and perhaps RLC programs in some cities*).

  Using the same aggregate-economic-effect analysis methodology described in the preceding bullet, it appears that warning signs located at both the treated intersection and at the city limits were associated with a larger benefit than warning signs at intersections only.

- *If other improvements are made during the installation of RLCs, what portion of the change in intersection crashes is due to these improvements and what portion is due to the RLC?*
  It was not possible to differentiate RLC effects from the effects of other changes at the time of treatment simply because there were virtually no sites where other changes were made at the same time according to information provided to the project team.
- *What effect does a "good" public information program have on safety at intersections with RLCs?*
- Again, using the same methodology, high publicity level was found to be associated with a greater benefit than medium publicity level.
- *What is the effectiveness of "fine only" (i.e., owner liability) program versus "fine and points" (i.e., driver liability) program? (Second-level priority.)*
  Again using the same methodology, "fine and points" was found to be associated with a greater benefit than "fine only."

In summary, the multijurisdiction database developed and the crash-based and economic analyses used made it possible to answer most of the questions posed by FHWA. This economic analysis represents the first attempt in the known literature to combine the positive effects of right-angle crash reductions with the negative effects of rear end crash increases, and to identify factors that might further enhance the effects of RLC systems. Larger crash sample sizes would have added even more information. The following primary conclusions are based on these current analyses:

- Even though the positive effects on right-angle crashes of RLC systems is partially offset by negative effects related to increases in rear end crashes, there is still a modest to moderate economic benefit of between \$39,000 and \$50,000 per treated site year, depending on whether one examines only injury crashes or includes PDOs, and on whether the statistically non-significant shift to slightly more severe right-angle crashes remaining after treatment is, in fact, real.
- Even if modest, this economic benefit is important. In many instances today, the RLC systems pay for themselves through red-light-running fines generated. However, in many jurisdictions, this differs from

most safety treatments where there are installation, maintenance, and other costs that must be weighed against the treatment benefits.

- The modest benefit per site is an average over all sites. As the analysis of factors that impact showed, this benefit can be increased through careful selection of the sites to be treated (e.g., sites with a high ratio of right-angle to rear end crashes as compared to other potential treatment sites) and program design (e.g., high publicity, signing at both intersections and jurisdiction limits).

The authors close with two additional findings from the study. First, safety studies related to local road systems, as opposed to State highway systems, will be more expensive and difficult to conduct, and this should be considered in the research planning process.

While most State DOTs have developed computerized databases containing crash, roadway inventory, and traffic data that can be used in research (including FHWA's multi-State Highway Safety Information System), this study has further documented that such files (particularly historical data other than crash data) are not usually available in local jurisdictions.

If available, the data are often not computerized, they are in multiple formats, and they are maintained by different offices; thus, jurisdiction choice is a critical part of such research planning, with the availability, types, and quality of data being key issues.

Second, research involving signalization parameters such as yellow interval and cycle length is difficult. The difficulties surface not only because historic signal-change data may be difficult to acquire, but the research may be made more difficult by adaptive signal systems where the parameters change within a day.

This is particularly true if traffic volumes by intersection approach cannot be linked to the specific sets of signal parameters; AADT is not usually sufficient. Research on these topics is important, but careful planning will be required to ensure its success.

APPENDIX A. SOURCES OF DATA BY JURISDICTION

Table 19. Data sources

Data Item	Baltimore, MD	Charlotte, NC	El Cajon, CA	Howard County, MD
Crash data	Maryland CODES data from National Study Center for Trauma and MS University of Maryland— Baltimore	Charlotte Department of Transportation (DOT)	California Highway Patrol Crash Database (SWITRS)	Maryland CODES data from National Study Center for Trauma and MS University of Maryland—Baltimore
Intersection traffic counts	From Baltimore Traffic Management Centers (TMCs), Maryland State Highway Administration counts, MPO (BMC) counts, and vendor loop data	Database of total intersection entering volumes, hardcopied TMCs, midblock counts, and approach volumes on signal plans	El Cajon turning movement counts and permanent station counts Link volumes from the area MPO (SANDAG)	Howard County TMCs and midblock ADTs Maryland State Highway Administration (MDSHA) TMCs and midblock ADTs
Number of approach lanes, medians, turn lanes	Aerial photos from Baltimore GIS data	Signalized: Most are from detailed signal plans, a few from aerial photographs Unsignalized: Estimated from aerial photos	Signalized: Signal plans Unsignalized: Aerial photos, signal warranting study reports, and drawings of adjacent intersections	Signalized: Howard County and MDSHA signal plans, aerial photos
Intersection "change" data	Baltimore historical timing sheets	No signal timing change data available.	El Cajon historical signal timing sheets	From historical signal plans and log maintained by Howard County staff

Table 19. (Continued)

Data Item	Baltimore, MD	Charlotte, NC	El Cajon, CA	Howard County, MD
Traffic signal phasing data	Baltimore printouts of current signal timing	Charlotte signal timing system	El Cajon printouts of current signal timing	Howard County and MDSHA signal plans
LEDs and backplates	No data available	From signal drawings	Used by El Cajon, but no formal installation dates were available	From log maintained by Howard County staff
RLC installation date	City of Baltimore Traffic Engineering Division	Charlotte DOT	El Cajon Traffic Engineering Division and Police	Howard County staff
RLC signing	City of Baltimore policy	Charlotte DOT policy	El Cajon Traffic Engineering Division policy	Howard County policy
RLC ticket protocol	City of Baltimore Traffic Engineering Division	Charlotte DOT	El Cajon Traffic Engineering Division	Howard County staff

Data item	Montgomery County, MD	San Diego, CA	San Francisco, CA
Crash data	Maryland CODES data from National Study Center for Trauma and MS University of Maryland—Baltimore	California Highway Patrol Crash Database (SWITRS)	California Highway Patrol Crash Database (SWITRS)
Intersection traffic counts	Montgomery County ADTs and TMCs	San Diego machine counts and turning movement counts	San Francisco ADT files, hardcopied turning movement counts and approach counts State-maintained ADTs

Data item	Montgomery County, MD	San Diego, CA	San Francisco, CA
Number of approach lanes, median presence, presence of turn lanes	Signalized: Signal plans supplemented with aerial photos Unsignalized: Intersection drawings on TMCs	Signalized: Signal plans (outdated), aerial photographs, non-scale drawings from report Unsignalized: Work-order requests, sketches on turning movement counts, and coders assumptions based on volume	Signal plans, pavement marking drawings, and aerial photographs
Intersection "change" data	Montgomery County historical signal timing files	San Diego historical timing files	San Francisco historical timing cards
Traffic signal phasing data	Montgomery County signal timing files	San Diego signal timing files	San Francisco signal timing files
LEDs and backplates	These data are not maintained	Requested but never received from City of San Diego maintenance staff	Although the city uses LEDs, it was not able to provide installation dates
RLC installation date	Montgomery County Department of Public Works (DPW)	City of San Diego Traffic Engineering Division	San Francisco Department of Parking and Traffic (DPT)
RLC signing	Montgomery County DPW provided general information	City of San Diego Traffic Engineering Division policy	San Francisco signing database
RLC ticket protocol	Montgomery County DPW	City of San Diego Traffic Engineering Division	San Francisco DPT

APPENDIX B. NUMBER OF TREATED, REFERENCE, AND COMPARISON SITES IN EACH JURISDICTION

Table 20. Sites by jurisdiction

Jurisdiction	Treated Sites	Signalized Reference and Spillover Analysis Sites	Unsignalized Comparison Sites
Baltimore, MD	19	86	46
Charlotte, NC	31	74	42
El Cajon, CA	6	53	38
Howard County, MD	18	34	38
Montgomery County, MD	21	55	40
San Diego, CA	19	54	44
San Francisco, CA	18	52	48
Total	132	408	296

REFERENCES

[1] Retting, R.A., R. Ulmer, and A. Williams. "Prevalence and characteristics of red-light running crashes in the United States." *Accident Analysis and Prevention* 31:687-94, 1999.

[2] Persaud, B., F. Council, C. Lyon, K. Eccles, and M. Griffith. "A multi-jurisdictional safety evaluation of red-light cameras." *Transportation Research Record.* Transportation Research Board, in press.

[3] Council, F., B. Persaud, C. Lyon, K. Eccles, M. Griffith, E. Zaloshnja, and T. Miller. "Guidance for implementing red-light camera programs based on an economic analysis of safety benefits." *Transportation Research Record.* Transportation Research Board, in press.

[4] Hauer, E., *Observational Before-After Studies in Road Safety: Estimating the Effect of Highway and Traffic Engineering Measures on Road Safety.* Pergamon Press, Elsevier Science Ltd., Oxford, United Kingdom. 1997.

[5] Zaloshnja, E., T. Miller, F. Council, and B. Persaud. "Comprehensive and Human Capital Crash Costs by Maximum Police-Reported Injury Severity within Selected Crash Types." Accepted for presentation at the 2004 Annual Meeting, American Association for Automotive Medicine, Key Biscayne, FL, September 2004.

[6] Council, F., E. Zaloshnja, T. Miller, and B. Persaud. *Crash Cost Estimates by Maximum Police-Reported Injury Severity within Selected Crash Geometries*. Federal Highway Administration HRT-05-051, U.S. Department of Transportation, Washington, DC. 2005.

[7] National Safety Council. (1990) Manual on Classification of Motor Vehicle Traffic Accidents, Fifth Edition (ANSI D-16.1-1989). Itasca, IL.

[8] Hillier, W., J. Ronczka, and F. Schnerring. *An Evaluation of Red Light Cameras in Sydney*. Report Number 1/93. Road Traffic Authority NSW: Road Safety Bureau. Rosebery, Australia. 1993.

[9] South, D., W. Harrison, I. Portans, and M. King. *Evaluation of the Red Light Camera Program and Owner Onus Legislation*. Victoria Transport: Road Traffic Authority. Hawthorn, Victoria. 1988.

[10] Andreassen, D. *A Long Term Study of Red Light Cameras and Accidents*. Research Report #261. Australian Road Research Board. Victoria, Australia. 1995.

[11] Kent, S., B. Corben, B. Fildes, and D. Dyte. *Red Light Running Behaviour at Red Light Camera and Control Intersections*. Report #73. Monash University: Accident Research Centre. Clayton, Australia. 1995.

[12] Mann, T., S. Brown, and C. Coxon. *Evaluation of the Effects of Installing Red Light Cameras at Selected Adelaide Intersections*. Walkerville, Australia: South Australian Department of Transport, Office of Road Safety, Report Number 7/94, 1994.

[13] London Accident Analysis Unit. *West London Speed Camera Demonstration Project: An Analysis of Accident and Casualty Data 36 Months "After" Implementation and Comparison with the 36 months "Before" Data*. London Research Centre: Environment and Transport Studies. London, England. 1997.

[14] Hooke, A., J. Knox, and D. Portas. *Cost Benefit Analysis of Traffic Light and Speed Cameras*. Police Research Series Paper 20. Police Research Group, Home Office. London, England. 1996.

[15] Ng, C.H., Y.D. Wong, and K.M. Lum. "The impact of red light surveillance cameras on road safety in Singapore," *Road and Transport Research* 6(2): 72-80, 1997.

[16] Retting, R. and S. Kyrychenko, *Crash Reductions Associated with Red Light Enforcement in Oxnard California*. Insurance Institute for Highway Safety, April 2001.

[17] *Safe*Light. Charlotte. Reviewed from data and reports on its website, http://www.ci.charlotte.nc.us/citransportation/programs/safelight.htm.

[18] Maryland House of Delegates Commerce and Government Matters Committee. "Automated Enforcement Review—Red-Light Running Detection Camera Systems, Howard County, MD." January 18, 2001.

[19] Fleck, J.L., and B.B. Smith. "Can we make red light runners stop? Red light photo enforcement in San Francisco, California," *Transportation Research Record 1693,* Transportation Research Board, 2001.

[20] Vinzant, Janet C., and B.J. Tatro. *Evaluation of the Effects of Photo Radar Speed and Red Light Camera Technologies on Motor Vehicle Crash Rates*. Prepared for the City of Mesa Police Department. March 1, 1999. Found at http://www.ci.mesa.az.us/police/traffic/march_1999_report.htm.

[21] Fox, Halcrow, *Accidents at Signal Controlled Junctions in Glasgow*. Glasgow, Scotland. The Scottish Office Central Research Unit, 1996. (Note that the review was based on the detailed summary posted on the Web site, not the full report.) Found at http://www.scotland.gov.uk/cru/resfinds/drf23-00.htm (accessed 2/25/05).

[22] Winn, Ray. *Running the Red and Evaluation of Strathclyde Police's Red Light Camera Initiative.* The Scottish Office Central Research Unit: Glasgow, Scotland. 1995. (Note that the review was based on the Web-based report rather than the full report.) Found at http://www.scotland.gov.uk/cru/resfinds/drf7-00.htm.

[23] Retting, R.A., S. Ferguson, and A.S. Hakkert. "Effects of red light cameras on violations and crashes: A review of the international literature." *Traffic Injury Prevention*, 4:17-23, 2003.

[24] Harwood, D. W., F.M. Council, E. Hauer, W.E. Hughes, and A. Vogt. *Prediction of the Expected Safety Performance of Rural Two-Lane Highways.* Federal Highway Administration RD-99-207, U.S. Department of Transportation, Washington, DC. 2000.

[25] Lord, D., *The Prediction of Accidents on Digital Networks: Characteristics and Issues Related to the Application of Accident Prediction Models*. Department of Civil Engineering, University of Toronto, Toronto, Canada. 2000.

[26] Blincoe, L., A. Seay, E. Zaloshnja, T. Miller, E. Romano, S. Luchter, and R. Spicer. *The Economic Impact of Motor Vehicle Crashes,* Washington, DC: U.S. DOT National Highway Traffic Safety Administration, May 2002.

[27] Miller, T., D. Lestina, M. Galbraith, T. Schlax, P. Mabery, and R. Deering. "United States Passenger-Vehicle Crashes by Crash Geometry:

Direct Costs and Other Losses." *Accident Analysis and Prevention,* Vol. 29, No. 3, May 1997.

[28] Hall, J.W., "Economic benefit of accident reductions," Proceedings of the 68th Annual Meeting of the Institute of Transportation Engineers, Washington, DC: ITE, 1998.

[29] Genstat 5 Committee, Statistics Department, Rothamsted Experimental Station. *Genstat 5 Release 3 Reference Manual.* R.W. Payne (committee chairman). Clarendon Press, Oxford, United Kingdom, 1993.

[30] National Highway Traffic Safety Administration. *National Accident Sampling System Crashworthiness Data System 1999–2001*. NHTSA: Washington, DC, 2002.

[31] National Highway Traffic Safety Administration. *National Accident Sampling System 1984–1986*. NHTSA: Washington, DC, 1987.

[32] National Highway Traffic Safety Administration. (NASS) General Estimates System (GES). NHTSA: Washington, DC. 2004. Found at <http://www-nrd.nhtsa.dot.gov/departments/nrd-30/ncsa/TextVer/GES.html>.

[33] Persaud, B.N., H. McGee, C. Lyon, and D. Lord. "Development of a procedure for estimating the expected safety effects of a contemplated traffic signal installation." *Transportation Research Record,* 1840, pp. 96–103, 2003.

OTHER REFERENCES

Breiman, L., R. Olshen, J. Freidman, and C. Stone. *Classification and Decision Trees*. Wadsworth International Group, Belmont, CA. 1984.

Butler, Pamela Crenshaw. *A Quantifiable Measure of Effectiveness of Red Light Running Cameras at Treatment and Non-Treatment Sites*, Washington, DC: Howard University Thesis, May 2001.

Feber, D. J., K.J. Crocker, and J.M. Feldmeier. "The AAA Michigan Road Improvement Demonstration Program: an Analysis of the Effectiveness of Using Safety Enhancements to Help Reduce Societal and Insurance Costs." Transportation Research Board, Washington, DC. 2000.

Hauer E., J. Lovell, and B.N. Persaud, *New Directions For Learning About Safety Effectiveness*. Federal Highway Administration and National Highway Traffic Safety Administration, Report FHWA/RD-86-015, Washington, DC. January 1986.

Hauer, E., and J. Bamfo. *Two Tools for Finding What Function Links the Dependent Variable to the Explanatory Variables*. International Cooperation on Theories and Concepts in Traffic Safety, Lund, Sweden. 1997.

Maisey, G. *The Effect of a Mechanical Surveillance Device on Urban Signalized Intersection Accidents*. Research and Statistics Report No. 17. Road and Traffic Authority, Western Australia. 1981.

Makinen, T., and H.L. Oei. *Automatic Enforcement of Speed and Red Light Violations: Applications, Experiences and Developments*. Report No. R-92-58, SWOV Institute for Road Safety Research, Leidschendam, Holland. 1992.

Persaud, B. N., R.A. Retting, P. Garder, and D. Lord. "Safety Effect of Roundabout Conversions in the U.S.: Empirical Bayes Observational Before-After Study." *Transportation Research Record: Journal of the Transportation Research Board,* No.1757, Transportation Research Board, National Research Council, Washington, DC. 2001.

Radalj, T., *Evaluation of Effectiveness of Red Light Camera Programme in Perth,* Main Roads Western Australia (an Australian Governmental agency). Found at http://www.monash.edu.au/cmo/roadsafety/abstracts_and_papers/026/Tony_Radalj_Paper_26_re vised.pdf (accessed 2/25/05).

Retting, R.A., J. Chapline, and A. Williams. *Changes in Crash Risk Following Re-Timing of Traffic Signal Change Intervals*. Insurance Institute for Highway Safety, September 2000.

Zaal, D. *Traffic Law Enforcement: A Review of the Literature*. Monash University Accident Research Centre, Report No. 53. Clayton, Australia. 1994.

In: Red Light Camera Systems
Editor: Cynthia Munoz
ISBN: 978-1-63483-448-3

Chapter 2

RED LIGHT CAMERA SYSTEMS: OPERATIONAL GUIDELINES*

Federal Highway Administration

FOREWORD

Red light running is one of the major causes of crashes, deaths, and injuries at signalized intersections. Most recent crash statistics show that nearly 1,000 Americans were killed and 176,000 were injured in 2003 due to red light running related crashes. The monetary impact of crashes to our society is approximately $14 billion annually.

The Federal Highway Administration (FHWA) and National Highway Traffic Safety Administration (NHTSA) support a comprehensive approach to intersection safety that incorporates engineering, education, and enforcement countermeasures to prevent red light running and improve intersection safety. Red light cameras can be a very effective countermeasure to prevent red light running. There are a number of studies that indicate reduction in crashes at signalized intersections due to red light cameras. FHWA is promoting red light cameras as one of its identified priority, market-ready safety technologies.

This document is an update to a previous version dated March 2003 [1]). The information contained in this document is intended to foster discussions

* This is an edited, reformatted and augmented version of a report, FHWA-SA-05-002, issued by the U.S. Department of Transportation, Federal Highway Administration, January 2005.

and initiatives that will improve intersection safety by reducing crashes due to red light running. This document is not a regulatory requirement and the decision to use red light cameras is a matter for local decision-makers.

SECTION I. INTRODUCTION

The use of camera systems for the enforcement of red light violations at signalized intersections is increasingly widespread in the United States. State and local agencies have found that the use of red light camera systems can reduce red light running by motorists and, more importantly, reduce the number of crashes attributable to red light violations. A recent synthesis of literature on the safety impacts of red light camera systems found that there was "...a preponderance of evidence, albeit not conclusive, indicating that red light running camera systems improve the overall safety of intersections where they are used... angle crashes are usually reduced, and, in some situations, rear-end crashes increase, but to a lesser extent."[2] Furthermore, a recent study of red light camera systems in seven jurisdictions throughout the US confirmed that these systems are likely to reduce right-angle crashes but can increase rear-end crashes. However, the systems were found to provide an economic benefit of $28,000 to $50,000 at a treated site when considering the economic cost of crashes by crash type [3]. The reduction in the number of crashes is especially important as crashes caused by motorists running red lights are, on the average, more deadly and damaging than other types of crashes at signalized intersections [4, 5, 6].

The purpose of these guidelines is to assist jurisdictions who are considering the implementation of red light camera systems and help them avoid inconsistent or incorrect application of such systems. Questions have been raised regarding the contracting, design, implementation, operation of red light camera systems, and the legality and intent of photo enforcement systems. In a broader perspective, for continued use of red light camera systems and other technologies to improve transportation operations and safety, it is vital these technologies are perceived as accurate and reliable and are applied fairly.

Red light cameras are currently in place in more than 95 communities in the United States. As red light camera programs continue to be implemented across the nation, there is much to be learned from previous experiences. Traffic engineers, law enforcement officials, and other State and local agency managers can benefit greatly from guidance and research that provides effective and comprehensive procedures for implementing a successful red light camera program.

The Federal Highway Administration (FHWA) and the National Highway Traffic Safety Administration (NHTSA) have developed this operational guideline for use by State and local agencies for the implementation and operation of red light camera systems. Although not a regulatory requirement, the guideline is intended to provide critical information for State and local agencies on relevant aspects of red light camera systems in order to promote consistency, proper implementation, and operation; and to ensure that this effective tool and other forms of technology remain available to transportation and enforcement agencies around the nation.

This guideline can be used by State and local agency managers, transportation engineers, and law enforcement officials to identify and properly address safety problems resulting from red light running within their jurisdiction.

This guideline outlines proven and effective practices implemented in the United States, and provides operational guidance that can be followed to ensure that cost-effective solutions are implemented by State and local agencies. The Institute of Transportation Engineers (ITE) has issued *Making Intersections Safer: A Toolbox of Factors and Countermeasures to Prevent Red Light Running* [7] that discusses in detail many of the issues and practices identified in this document.

SECTION II. UNDERSTANDING OF THE PROBLEM

Red light running and the collisions and injuries that result from it has become a national safety problem. Incidences of red light running, along with other aggressive driving behaviors, is on the rise. For 2003, the most recent year for which statistics are available, there were 206,000 red light running crashes, resulting in 934 fatalities and 176,000 injuries [8].

Factors Contributing to Crashes Caused by Red Light Running

A number of factors that contribute to crashes caused by red light running have been identified from research and crash data studies. These factors include namely:

- Driver behavior.
- Intersection design and operation.
- Vehicle characteristics.
- Weather.

Driver Behavior

Driver behavior (including speeding and aggressive driving) is the most significant contributing factor to the occurrence of red light running. Motorists may accelerate when anticipating a change in signal indication, in order to make it through the intersection on the yellow. If a motorist misjudges the time of the signal change, he or she will enter the intersection against the red signal indication.

Motorists driving above the posted speed limit or driving too fast for conditions increase the distance needed to stop before entering intersections and decrease the distance available to react to a change in traffic signal indication. In other words, speeding significantly increases the risk of running a red light and the possibility of being involved in a crash as a result of running the red light.

For many drivers, inattentiveness may also be the contributing cause. Drowsiness, conversing with passengers, eating, and use of a cellular phone or other electronic devices are among the many common distractions that cause drivers to reduce their focus on the task of driving. Inattentive or distracted drivers may perceive a change in signal indication late or in some cases not at all. Motorists, perceiving the signal indication late, may not perceive the change in time and run through the red signal.

Intersection Design and Operation

Deficiencies in the design and configuration of signalized intersections may contribute to red light violations. Certain design and configuration conditions (geometrics) may provide inadequate stopping distance, may cause motorist to be confused, or limit visibility of traffic control devices. Where these design and configuration conditions can be identified and corrected through engineering improvements, the number of red light violations can be reduced. Therefore, it is important when considering the use of a red light camera system, that an engineering study be done to identify potential engineering improvements that could be implemented in the intersection design and configuration. If an engineering study identifies countermeasures that might take considerable time to design and implement, then appropriate short-term temporary solutions should be considered while long-term improvements are investigated and implemented.

An engineering study can identify the following conditions that may be present at a signalized intersection and contribute to red light running by motorists:

- *Grade.* The grade of an intersection approach may significantly effect the time and distance needed for a motorist to stop a vehicle at an intersection. If approaching the intersection on a downhill grade, motorists may not account for vehicle mass and momentum, which will require longer stopping time [9].
- *Poor Visibility.* Poor visibility due to darkness, rain, or snow and to a motorist's impaired vision may prevent or reduce a motorist's ability to see and react to signs, signals, and other traffic control devices at intersections and on the approaches to intersections in a timely manner.
- *Roadside Obstructions.* Roadside obstructions (i.e., parked vehicles, vegetation growth, pedestrians) may block a motorist's field of view to road signs, traffic signals, and other features at the intersection, thus contributing to driver confusion. Intersections and adjoining approaches should be engineered so that roadside parking does not interfere with sign visibility. Signs should be regularly monitored and cleared of vegetation over-growth.
- *Line of Sight.* As motorists approach an intersection, their line of sight to the intersection should be unobstructed. Any obstructions may reduce reaction times and/or negatively impact driver behavior. Line of sight problems often occur at intersections located at the base of a hill, where the traffic signal is partially or completely hidden from the driver's view until reaching the top. Line of sight problems may also occur when following a taller vehicle, whereby the taller vehicle obstructs the line of sight of the driver of the following shorter vehicle.
- *Traffic Volumes.* Research studies have indicated that time of day and traffic volume may be associated with increased red light running behavior [5, 10]. During peak traffic periods, increased traffic volumes and congestion may contribute to the number of motorists

running red lights. Motorists traveling during these periods are often subject to the delays from traffic congestion that may negatively affect their driving behavior. If traffic signals are not properly timed to accommodate the increased traffic volumes and coordinated to enhance traffic flows, motorists may wait for two or more cycles before passing through signalized intersections. Frustrated motorists may choose to enter the intersection on a red light in order to avoid waiting through an additional cycle. Traffic volumes during the late evening and early morning hours are relatively light. During these time periods and especially on non-traffic-actuated approaches at signalized intersections, motorists may have to wait for extended times when there is little or no traffic on the other intersection approaches. Faced with this situation, motorists may engage in unsafe or aggressive driving behavior by electing to run the red light. A study has also indicated that larger intersections and high volumes seem to be related to red light running [10].

- Signal Timing. Research studies have determined that inadequate signal timing generally tends to increase red light running by motorists at signalized intersections [9, 11, 12]. Methods for development of signal timing plans and discussion of adequate clearance intervals can be found in the Traffic Control Systems Handbook [13] and the Manual of Traffic Signal Design [14] available from the ITE (www.ite.org).

Vehicle Characteristics

Vehicle characteristics may contribute to red light running and to crashes resulting from red light running. Vehicles that carry heavy loads require additional time to slow and stop when a traffic signal changes to yellow [15]. Drivers of vehicles with heavy loads may forget or disregard the effect of the loads on stopping distances, and this may result in red light running.

Weather

One study has found that weather is not a predictor of red light running [10]. However, it is reasonable to infer that weather conditions such as heavy rain, snow, hail, or high winds may distract drivers, make roadway surfaces slick, and may cause stopping distance to be increased. Inclement weather conditions will likely exacerbate the effects of steep grades, limited sight distances, and high approach speeds. Reduced visibility resulting from severe weather, sun glare, or dust and debris may also prevent a motorist from observing signs, signals, or other traffic control devices in a timely manner. Location and configuration of signals relative to early morning and afternoon sun glare can reduce visibility of signal colors.

SECTION III. PROBLEM IDENTIFICATION

An initial step in determining if red light camera systems, or any other countermeasure, are to be employed, is to establish if a red light running and resulting crash problem exists in the jurisdiction in general or at a specific intersection. Red light running and crashes attributable to red light running may result from a number of contributing factors and, consequently, may be addressed by a variety of countermeasures encompassing engineering improvements, enhanced driver and public education, and increased enforcement. The red light running problem at any intersection needs to be investigated and the feasibility of all countermeasures, including red light camera systems, should be addressed.

Investigating Intersection Safety

A systematic approach to the collection and analysis of various intersection safety-related data is important for the identification of intersections where there is a high incidence of red light running and for the investigation of countermeasures. The elements of the investigation are described below.

Data Collection

An identification of intersections with high crash rates, public complaints, and those identified by law enforcement as having violation problems, is the first step in improving intersection safety. While complaints or other inputs from motorists and the general public about red light running at specific locations are helpful, data on crashes resulting from red light running and the number of red light violations at signalized intersections are required for an objective assessment of the potential safety problems and an understanding of factors that may be contributing to the problems.

Data for investigating intersection safety may be obtained from the following sources:

- Crash statistics and investigation records maintained by law enforcement and traffic engineering agencies.
- Crash statistics maintained by insurance companies, if available.
- Counts of citations issued by law enforcement officers for red light running.
- Camera surveys of driver behavior at intersections, including counts of red light violations.
- Field observations of driver behavior at intersections, including speed surveys, by trained personnel.
- Complaints or other inputs from motorists and the general public.

Intersection Crash Data

State and local agencies typically collect crash data for injury crashes or crashes where property damage exceeds a pre-determined threshold amount. Generally, data regarding minor non-injury crashes are not collected.

Crash data is the most comprehensive basis for the identification and analysis of red light running at signalized intersections. The data should be classified with as much detail as possible, including:

- Intersection location, by identifier or street names.
- Crash type, for example, angle collision, rear-end collision, or striking a pedestrian or bicyclist.
- Crash location and vehicle movement, including movement direction, left turn, through, or u-turn.
- Crash day of week and time of day.
- Weather at the time of the crash.
- Type of vehicle: automobile, SUV, truck, bus, or motorcycle.
- Vehicle speeds.

Red Light Violation Data

An analysis of red light running violation data may serve as an alternative to the crash data when crash data are not available. This data may be available from records maintained by law enforcement or from special studies collected by video cameras or other means. However, this approach is generally not recommended because the data may reflect targeted enforcement at selected intersections only and the criteria applied by individual officers for issuing citations may vary from one officer to another.

Driver Behavior Observations

Video surveys or field observations may also provide important data on driver behavior and operational conditions at intersections. Conditions such as traffic repeatedly backing up into an intersection from adjacent freeway ramps or intersections, traffic backing into the through lane from the left-turn bay, or high speeds on the intersection approaches may be contributing to red light violations and to the incidence of crashes attributable to red light running.

Traffic, Signal, and Intersection-Related Data

Intersection geometry, traffic volumes, and signal timing data are generally available from the State or local agency traffic engineering or public works department. The following data provide necessary information for a rigorous analysis of factors that may be affecting the frequency of crashes attributable to red light running:

- Signal operation: coordinated, fully actuated, semi-actuated, pre-timed, or isolated.
- Signal phasing: protected, permissive, or protected-permissive left turns, split phasing.
- Yellow interval.
- All-Red interval.
- Number of traffic lanes.
- Number and location of signal heads.
- Vehicle detector locations.

- Approach angles, speed limits, and directions.
- Street lighting.
- Approach grades and visual obstructions.
- Average daily and peak period traffic volumes.
- Proportion of tall or wide vehicles.

A sample assessment form that can be used as a guide for field inspections of problem sites can be found in *Intersection Safety Issue Briefs* [16].

Motorist Complaints and Comments

Qualitative means to identify intersections where red light running is a frequent occurrence should also be employed by State and local agencies. Through the solicitation of inputs from motorists and from the general public, intersections where there are unique or changed conditions or where motorists have witnessed "near misses" that might not otherwise be identified from an analysis of crash data can be considered for further analysis and investigations. Written and oral complaints from motorists may be used as an input for determining intersections where there is a problem with red light running.

Engineering Study

The State or local agency considering the use of a red light camera system should conduct an engineering study to determine the factors contributing to red light running and to identify appropriate countermeasures that could be implemented to reduce the number of crashes resulting from red light violations. Once identified, the appropriate effective countermeasures (engineering, education, and enforcement) should be considered in addressing the crash problem.

Section 4C.01 of the *Manual on Uniform Traffic Control Devices (MUTCD)* requires that an engineering study be conducted whenever the installation of a traffic signal is being considered [17]. After a traffic signal is installed, traffic conditions may change and a high incidence of red light violations may occur, a changed condition that warrants attention by traffic engineers and traffic safety professionals. Engineering studies should be fully documented in preparation for any questions or concerns about proposed photo enforcement camera installation. The documentation should include a full description of the operation of the intersection, assessed throughout the day. The engineering study must also review pedestrian and bicycle conditions at the intersection. For further discussion on the engineering studies, refer to the MUTCD at http://mutcd.fhwa.dot.gov/. For a complete description of the steps in an engineering study, please refer to ITE's *Manual of Transportation Engineering Studies [18]*.

SECTION IV. COUNTERMEASURES AND THEIR APPLICATIONS

Over the last decade, considerable efforts have been made to mitigate red light running behavior by motorists in the United States. Research has shown that engineering improvements [9, 11, 19], safety education and increased enforcement by law enforcement officers [12, 20] can significantly reduce red light violations. In addition, to supplement traditional law enforcement activities, many jurisdictions have implemented automated enforcement red light camera systems.

The solution to the problem of red light running and resulting crashes may require one or a combination of engineering, education, and enforcement measures.

- *Intersection Engineering Improvements.* Engineering solutions to be considered include, but are not limited to, modifying traffic signal timing, improving signing and marking, improving sight lines, modifying grades and/or grade separation, adjusting the prevailing speeds, changes in surface treatments, altering lane configuration, and replacing the traffic signal with some other form of traffic control device or intersection type.
- *Education.* A well-designed public information and education campaign will assist motorists and the general public in understanding the safety issues inherent to red light running. It will provide information and data that explain what red light running is, why red light running is dangerous, and what actions are currently being undertaken to reduce the incidence of red light running.
- *Traditional Enforcement By Law Enforcement Officers.* Traditional enforcement efforts by law enforcement officers specifically targeting

red light running violators can be a cost effective deterrent in reducing red light violations at problem intersections.

- *Red Light Camera Systems.* Red light camera systems can be a cost effective tool to reduce red light violations. Red light camera systems should be part of a comprehensive intersection safety program, which considers all countermeasures to reduce fatal and injury crashes at intersections.

An engineering study should consider each of these possible solutions in order to identify the most appropriate solution to the documented problem at the intersection.

Engineering Countermeasures

Intersection design deficiencies may contribute to red light running and crashes at signalized intersections. The deficiencies may be mitigated by engineering improvements of two types: traffic operation (including signal control) improvements, and intersection geometry improvements.

Traffic Operation and Signal Control

At a minimum, retiming of the traffic signal should be analyzed as a red light running countermeasure. Signal timing should be reviewed regularly to determine if it is still appropriate for the traffic conditions in effect, and changed if the need for a change is indicated. Traffic signal timing, especially the length of the yellow and all-red interval times, should be in accordance with the broad guidelines in the MUTCD [17] and due consideration should be given to the informational report developed by ITE [9], which discusses methods for determining vehicle signal change and clearance intervals. In addition, any applicable State and local agency policies and procedures should also be followed.

The following list identifies possible engineering countermeasures to reduce incidences of red light running:

- *Improving Signal Head Visibility.* Signal head visibility can be improved by increasing the size of the traffic signal lamps from 8 to

12 inches. Improving signal visibility can be especially beneficial on streets that run in an east-to-west direction where the sun angle silhouettes the traffic signal head making it difficult to see the signal indication. The addition of backplates can also make signals more visible.

- *Additional Signal Heads.* Depending on the intersection and the number and visibility of signals currently deployed, adding signal heads may help decrease the frequency of red light violations. If a single signal head is used for multiple lanes, such as two through lanes, the signal may be blocked from view of a motorist if traveling behind or along side a truck or other high profile vehicle. A similar situation may occur when the traffic signal pole and head are located on the corner of an intersection.
- *All-Red Interval.* An all-red clearance interval provides additional time for motorists already in the intersection to proceed through the intersection on the red indication while holding cross traffic on the cross street approaches. The red clearance interval is not intended to reduce the incidence of red light running; rather it is a safety measure. The MUTCD indicates that the length of the all-red interval should be a function of traffic speed, cross street width, and length of the yellow interval [17]. The MUTCD guidance is that the clearance interval should not exceed six seconds in length. Typically where used, the length of an all-red interval is one second to not more than three seconds.
- *Appropriate Yellow Times.* The purpose of the yellow interval is to warn approaching traffic of the imminent change in the assignment of right-of-way. The length of the yellow interval is determined in such a way that it provides enough time for a vehicle to travel at its prevailing speed through the intersection before the traffic signal turns red or to allow a driver to stop at a comfortable average deceleration before entering the intersection. Therefore, the likelihood of a motorist running a red light increases as the yellow interval is shortened. Lengthening the yellow interval, within appropriate guidelines, has been shown to significantly reduce the number of inadvertent red light violations [11]. The length of the yellow intervals should be in accordance with MUTCD guidelines [17] and applicable State and local agency policies and procedures. The ITE informational report [9] contains more detailed discussion of methods for the calculation of appropriate clearance intervals for specific circumstances.

- *Signalized Intersection Warning Signs.* Advance warning signs are posted upstream on an intersection approach to alert motorists that they are nearing a signalized intersection. Advance warning signs are especially beneficial at intersections with curved approaches or those with steep grades.
- *Advanced Yellow Flashing Lights.* Consideration should be given to the use of advanced yellow flashing lights as advance warning at intersections with high-speed approaches or limited sight distances. These traffic control devices are posted well in advance of an intersection and only flash at approaching motorists when the signal indication is likely to be red when the motorists reach the intersection. This operation is different than the typical flashing yellow light in advance of an intersection that simply warns of the existence of the signalized intersection. Advanced yellow flashing lights may provide the most benefit in slowing vehicles on steep grade approaches and larger vehicles with more mass and momentum. Advanced warning flashers and their effect on red light violations were studied in Bloomington, Minnesota [21]. The advanced warning flashers were used for approximately three months, during which red light running violation data was collected. It was determined that the installation of the advanced yellow flashing lights reduced red light violations significantly at the study intersection, with a greater reduction in truck red light violations.
- *Adjusting the Approach Speed.* Approach speeds are a critical determinant for the length of the yellow time at a signalized intersection. Speed limits on the approaches to a signalized intersection where there is a problem with red light running should be evaluated based on speed studies and observations. It may be necessary to consider additional speed-affecting measures in order to achieve the necessary result.
- *Traffic Signal Coordination.* A coordinated traffic signal operation where motorists are able to move smoothly in platoons from intersection to intersection reduces the risk of red light violations and collisions.
- *Advance Vehicle Detection.* Advance vehicle detection may be employed to hold green signal indications for the maximum allowable time, allowing motorists at the back of platoons or under light traffic conditions, to legally enter and traverse a signalized intersection.

- *Removal of Unwarranted Traffic Signals.* Low volume, signalized intersections may experience a reduction in red light violations and crashes when traffic signals are removed and alternative intersection designs or other forms of traffic control are implemented. Guidance on the signal removal process can be found in Section 4B.02 of the MUTCD [17].
- *Removal of On-Street Parking.* The restriction of on-street parking for a distance of at least 200 feet from each intersection approach may enhance the driver visibility of signage, signals, pedestrians, cross-traffic, and other pertinent features near the intersection.

Other factors may serve to increase vehicle delays at signalized intersections and contribute to an increased frequency of red light running. Malfunctioning vehicle detection and signal actuating equipment, long phases or cycle lengths, or the use of protected left turn phases at times of the day when not required may serve to increase delays and, under certain circumstances, may cause motorists to engage in unsafe driving behavior by running red lights.

Intersection Geometry Changes

Deficiencies in intersection configuration or geometry may increase the number of motorists unintentionally running red lights. Where intersection geometry deficiencies are identified and can be improved, appropriate changes should be designed, deployed, and evaluated. Design guidelines can be found in AASHTO's *A Policy on Geometric Design of Highways and Streets*, 2001 [22] and ITE publications: *The Traffic Safety Toolbox: A Primer on Traffic Safety* [23], *Traffic Engineering Handbook* [24], and *Toolbox on Intersection Safety and Design* [25].

Education

A well-designed public information and education campaign will assist motorists and the public in understanding the safety issues inherent to red light running. This campaign should provide information and data that defines the red light running problem, explains why red light running is dangerous, and identifies the actions that are currently being undertaken to reduce the incidence of red light running. One of the key messages for the red light running education campaign should be the fatality and injury consequences and resulting emotional and economic toll of red light running. The emotional toll of red light running to crash victims and their families is quite obvious; however, the indirect economic costs associated with red light running related crashes in terms of lost productivity, higher insurance premiums, and medical cost, while significant, are often not understood.

An on-going educational program should be designed to combat red light running, in general, and be delivered in a way so as to communicate the seriousness of the violation and the effectiveness of the countermeasures being employed.

The on-going public information and education program should use various media, such as: posters, mailings, hand-outs, public service announcements on radio and television, warning notices, billboards, warning signs, press releases, slogans, and bumper stickers. The State or local agency should monitor the effectiveness of the educational program and modify it in order to achieve maximum effectiveness. A red light running education campaign supported by targeted enforcement by law enforcement agencies is a very effective tool. Red light running campaigns should be dovetailed with other traffic safety education and enforcement programs, such as speeding and other forms of aggressive driving.

Enforcement

Law enforcement officers play an important role enforcing traffic laws and rules of the road violations, which includes red light violations. Red light camera systems are but one method of monitoring and enforcing red light violations, others involve the direct use of law enforcement officers. Alternative officer enforcement strategies include:

- Single Officer: to enforce red light violations, an officer takes an inconspicuous position at an intersection where the officer can clearly see the signal and motorist. After observing a violation, it may be necessary to follow the violator through the intersection in order to stop and cite the driver.
- Dual Officer: a safer alternative, requiring a higher level of staff commitment, involves the use of separate observer and pursuit officers. The observer officer witnesses the violation and then radios the

information to the pursuit officer who is typically located downstream of the signal and will stop the driver and issue the citation.

- Multiple Signal Head Enforcement: a traffic signal head or some other lighted device is attached to the reverse side of an existing traffic signal. This allows a single police officer to observe violations from the opposite side of the intersection and to enforce red light violations in greater safety.
- Random Enforcement: refers to the random selection of the locations to be enforced and this may be performed by either single or multiple officers. Random enforcement makes police presence visible and reminds drivers that enforcement is taking place.
- Targeted Enforcement: is when problem locations are identified and officer staff resources are committed to enforcement for a particular period. Such stepped up enforcement can work as a visible reminder to motorists that the traffic laws should not be violated.

SECTION V. RED LIGHT CAMERA PROGRAM IMPLEMENTATION

A red light camera system is one of the measures available to traffic engineering, enforcement, and safety professionals, that when properly applied, may be effective in the reduction of certain types of collisions at signalized intersections. Red light camera systems have had the greatest success and highest levels of support in communities where they have been implemented as one element of an overall traffic safety management program. There are several key steps to successfully implementing a red light camera system program, which is the subject of this section.

Early Planning and Startup

The development of a successful red light camera program will be based on the systematic analysis of crash data, together with data on citations issued to motorists for red light running, where available, and inputs from the general public. The objective is to identify locations where red light running by motorists is contributing to crashes.

The key elements recommended for the early planning and startup of a red light camera program are as follows:

- Establish a Steering Committee.
- Establish Program Objectives.
- Identify the Legal Requirements.
- Assess System Procurement Alternatives.
- Establish Public Awareness and Information Campaign.

Steering Committee of Stakeholder Group Representatives

Any community considering the implementation of a red light camera system should first establish a steering committee inclusive of all stakeholders.

The Steering Committee serves to establish broad based program objectives and to monitor program results. The appropriate participants will vary by community and would typically include representatives from the following organizations:

- State Department of Motor Vehicles.
- State and local Police and Sheriff's Department.
- Traffic Engineering Department.
- Public Works Department.
- City, County, or State's Attorney's Office.
- City, County, or State Public Information Office or Community Affairs.
- Judiciary.
- Photo Enforcement Services Contractor, if one is hired.
- Selected Community Representatives.
- Selected outside Agency Representatives, such as a local Automobile Club.

A high level of quality control and on-going coordination of activities is required for the operation and maintenance of photo enforcement systems. The program also has significant visibility with the community at large and with their elected officials that require coordination to effectively communicate the program's objectives and program results.

Red Light Camera Program Objectives

Early on the Steering Committee should define as clearly as possible the red light camera program objectives. While it is clear that the overall objective of any red light camera program is the reduction of collisions at signalized intersections resulting from red light running, program objectives should address specific operational needs.

Legal Requirements

Prior to initiating a red light camera program, legal aspects and requirements should be identified. Red light camera systems pose legal questions and concerns, the answers to which may vary from State to State. In particular, privacy, citation distribution, and types of penalties need to be thoroughly addressed and resolved prior to the startup of a red light camera program.

Presently, there are two approaches that have been adopted by States in the deployment and operation of red light camera systems:

- Driver Responsibility. Where the government entity alleges that a driver has committed a violation and receives a citation, there should be photographic evidence that allows the driver to be identified. This requires that one or more red light camera(s) is/are located so that a frontal view of the vehicle is recorded as it runs the red light. Further, the recorded view should allow the driver and vehicle identities to be clearly determined. If the recorded view of a driver is obstructed or not clear, no citation should be issued. Additionally, a method should be provided through which the registered owner can certify that he or she was not the driver at the time of the violation. In States where red light camera systems are applied as described above, red light violations recorded by red light camera systems are considered to be moving violations with citations carrying the same penalties as citations issued by law enforcement officers, including "points" and holds on vehicle registration or driver license renewals for unpaid fines.
- Registered Owner Responsibility. Where the registered owner is responsible for the citation, only photographic evidence that identifies the vehicle, usually from the rear, and its license number is required.

> Typically, States where red light camera systems have been adopted in this manner have enacted legislation at the State level that authorizes the use of red light camera systems or permits local agencies to enact local ordinances for use of red light camera systems.

The National Committee on Uniform Traffic Laws and Ordinances (NCUTLO) developed the "Automated Traffic Law Enforcement Model Law" [26] to offer clear guidance to States considering automated enforcement technology.

Issues arising from legal challenges to automated photo enforcement are presented in Appendix A.

System Procurement Alternatives

There are a number of alternatives available to State and local agencies for the development and operation of red light camera programs. A State or local agency may take full responsibility for system operations and citation processing functions or elect to outsource these functions to a private contractor. Where a private contractor is responsible for installation and operation of the red light camera equipment, the State or local agency should establish the necessary procedures so that the agency has complete oversight and day-to-day supervision of the program. Table 1 summarizes selected alternatives for the acquisition, installation, operation, and maintenance of red light camera systems that are available to State and local agencies.

Where a private contractor is responsible for the processing of citations, compensation to private vendors based on the number of citations issued should be avoided. In multiple jurisdictions, the courts have determined that it is inappropriate for the private contractor to be responsible for determining installation locations and operation of the system because of an appearance of a conflict of interest. This conflict of interest should be avoided in all phases of the system installation and operation: startup, design, installation, operation, and maintenance. At all times, the State or local agency should verify and exercise complete oversight of all actions of the private contractor.

Some agencies are compensating their camera system vendors based on a flat fee per location per time period. Others have installed and operated their own systems. It may also be appropriate to pay a vendor to operate and maintain an agency-designed and -implemented system. Compensation should be based solely on the value of the equipment or the services provided.

Table 1. Selected Red Light Camera System Acquisition, Installation, Operation, and Maintenance Alternatives

	Project Planning and Management	Equipment Ownership	Design and Installation	Plan Check and Installation Inspection	Operation and Maintenance	Citation Data Processing	Decision To Issue Citation	Violator Inquires	Public Information Program
OPTION A									
State/Local Agency	☐			☐			☐		
Private Contractor		☐	☐		☐	☐		☐	☐
OPTION B									
State/Local Agency	☐	☐		☐			☐		
Private Contractor			☐		☐	☐		☐	☐
OPTION C									
State/Local Agency	☐	☐	☐	☐			☐	☐	☐
Private Contractor			☐		☐	☐			
OPTION D									
State/Local Agency	☐	☐	☐	☐	☐	☐	☐	☐	☐
Private Contractor									

Tables 2 and 3 summarize selected contractor payment options.

Public Awareness and Information Campaign

Education on improving traffic safety is a crucial component for any significant change to occur with traffic control systems. Appropriate educational elements should be applied regardless of the chosen solution. For red light camera programs, often the initial educational program includes issuance of warning citations to likely violators for limited period, and clear public communication of the date on which warning citations will be halted and actual enforcement citations will begin. However, education and media outreach efforts should continue throughout the life of the program to keep the public informed of results and need for safety vigilance. Ongoing awareness of the presence of enforcement measures is key to deterrence and longterm behavior changes.

Table 2. Payment Options for Contractor Owned and Operated Red Light Camera Systems

Payment Option	Equipment	Equipment Installation	Equipment Maintenance	Citation Data Processing
Initial Fixed Price Payment	☐	☐		
Initial Fixed Price Payment and Fixed Monthly Payments	☐	☐	☐	☐
Fixed Monthly Payments	☐	☐	☐	☐
Initial Fixed Price Payment and Per Citation Payments				
Per Citation Payments				
Initial Fixed Price Payment and Fixed Monthly Payment Schedule, Depending On Pre-Determined Low/High Number of Citations Issued	☐	☐	☐	☐
Fixed Monthly Payment Schedule, Depending On Pre-Determined Low/High Number of Citations Issued		☐	☐	☐
Time Worked and Materials Used			☐	☐

Table 3. Payment Options for Agency Owned and Contractor Operated Red Light Camera Systems

Payment Option	Equipment Maintenance	Citation Data Processing
Fixed Monthly Payments	□	□
Fixed Monthly/Per Citation Payments	□	□
Per Citation Payments		
Fixed Monthly Payment Schedule, Depending On Pre-Determined Low/High Number of Citations Issued	□	□
Time Worked and Materials Used	□	□

A red light camera program should not be started without a comprehensive public awareness and information campaign. Research has indicated that public information campaigns are a key to the success of the red light camera programs [27].

In 1995, FHWA sponsored a study examining the public's awareness of community-based safety programs. The study concluded that an information campaign needed to accomplish three objectives in connection with the implementation of red light camera programs. First, public awareness and information should make citizens more aware of their driving habits and safety consequences of running red light. This should stimulate a voluntary change in behavior at signalized intersections. Second, communications should be through a variety of media with the public and elected officials to explain program objectives, as well as program results. This is critical to gain public support for program expansion. Lastly, public awareness and information should provide motorists with advance warning that there is increased enforcement. This, by itself, may cause a change in driver behavior, but should describe the effectiveness of the systems. Without an effective educational campaign, motorists may be surprised or confused when they receive a citation. If questions or concerns can be effectively answered through written, telephone, or web-based information, motorists receiving citations will be more supportive of the program and less likely to question the program's overall objectives.

The public awareness and information campaign should encompass the following elements:

- Clear description of the operation of the red light camera equipment in non-technical terms.
- Clear statement of the program objectives.
- Description of the advantages of automated enforcement.
- Explanation of other measures being taken to improve safety at intersections.
- Description of the use of the red light camera program revenues.

The public awareness and information campaign may be developed using the following methods:

- Outreach efforts to employers, schools, driver education, local community groups, and all area media.
- Telephone and web-based information centers that include a hot-line for calls about intersection problems and traffic safety concerns, in addition to handling inquires regarding the operation of the red light camera program.

Public awareness and information campaigns are frequently used prior to and during the development of a red light camera program. The campaigns often employ a variety of methods in an effort to reach as many citizens as possible. The extent of the campaigns, however, varies among the jurisdictions where red light camera systems have been deployed. Table 4 identifies some of the more commonly used methods to increase public awareness and provide information.

An important aspect of the public awareness and information campaign is the direction provided for individuals who received citations on how to review their citation and/or view the photographic evidence.

It is also important for the success of the red light camera program that traffic court officials, including judges, commissioners, and administrative support personnel, be fully informed about the program scope and operation. Officials who often conduct traffic court hearings may not be fully versed in the operation of the red light camera equipment. It is important that the appropriate documentation is prepared and submitted in a timely manner in the event an individual contests the citation in court. The increased use of electronic data transfers and viewing may be appropriate to ensure that the court packages are readily available when needed.

Table 4. Public Awareness and Education Campaign Elements Used by Selected Red Light Camera Programs

Jurisdiction	Posters	Mailings	Hand outs	Media	Warning Notices	Billboards	Warning Signs	Press Releases	Slogans	Bumper Stickers
Charlotte, NC	■	■	■	■	■	■	■	■	■	■
Fairfax, VA		■		■	■		■	■		
Howard County, MD	■		■	■	■		■			
Lincoln, NE				■		■				
New York City, NY				■	■		■			
Oxnard, CA	■		■	■					■	■
Polk County, FL					■		■			
Sacramento, CA	■		■	■	■	■	■	■	■	■
San Francisco, CA				■	■	■	■	■		■
San Diego, CA		■		■			■	■		

Public awareness and educational outreach efforts for employers, schools, driver education programs, and local community groups, as well as the media, are necessary. Reports of program results, emphasizing the achieved safety benefits, should be available and posted on the program web site and local newspapers. The campaign should employ various communications media designed to reach residents and commuters, including regular surveys to gauge public support and awareness, and should focus on a central message of improving traffic safety. An example of a safety message is to emphasize that red light camera systems can be applied as an effective tool to reduce collisions resulting from red light running.

SYSTEM PLANNING

Proper planning by a State or local agency will establish the foundation for a successful red light camera system for detecting and documenting red light running at signalized intersections. As appropriate, a State or local agency should solicit assistance from other public agencies where red light camera programs have been successfully deployed, as well as from qualified consulting engineers with experience in red light camera systems design and operations.

Violations Processing Procedure

The violation processing procedure should address the following aspects of the installation and operation of the red light camera system, and the processing of the recorded violations and citations issued:

- Establish the enforcement threshold consistent with traditional enforcement methods.
- The number of days allowable from the date of the violation occurrence before citations can be mailed, if different from applicable legal requirements.
- How citations for commercial or rental car vehicles will be addressed.
- Minimum vehicle speed threshold.
- Should citation issuance be restricted to specific time periods or days of week only?
- Maximum number of days before citations are reissued to violators following registered owners disputed responsibility and subsequent violator identification.
- Guidelines for pitch measurement where inductive loops are employed for vehicle detection.
- Clear specification of photographic data requirements for issuing citations, including the red signal indication and the time elapsed since onset of red.

The system design and installation should be consistent with the definition of a violation under the applicable State and/or local laws.

The installation should be consistent with other neighboring intersections under the jurisdiction of the responsible agency, so that vehicle operators are held to a uniform standard throughout the jurisdiction.

Site Selection

Sites selected for the installation of red light camera systems should be based on accurate crash and red light violations data. As discussed earlier, data regarding the total number of crashes may be used, although intersections with high numbers of collisions may not have a high number of crashes related to red light running. Violation data needs to be applied with some caution. Likewise, locations where it is known that there are high numbers of red light violations may not have corresponding high numbers of crashes related to the red light running. Heavily traveled intersections where there are heavy left turn movements operated on protected left turn phases are often intersections of this type. Traffic volumes, except when used as a factor to determine the incidence of crashes or violations, are not a suitable measure for selecting locations for the installation of red light camera systems.

The installation of a red light camera system at a signalized intersection identified as having a red light running problem should be done when an engineering study of the intersection determines photo enforcement is an appropriate countermeasure to reduce the incidence of red light running.

Other criteria for red light camera system site selection may include recommendations from law enforcement and traffic safety professionals, citizens' complaints, and input from community groups. These criteria should be considered in conjunction with crash data and violations or citations data.

Undesirable characteristics that will also affect decisions regarding the installation of red light camera systems include:

- Driveways that restrict camera pole or auxiliary flash placement.
- Approaches that are more than three lanes wide and double left turn lanes where views are more frequently obstructed.
- Wide crossing streets where second photographs may not be taken at the predetermined location due to motorists speeding up and slowing down as they traverse the intersection.

When red light camera systems are in operation, law enforcement officials should place an emphasis on routine enforcement of traffic laws and regulations that require visible and unobstructed display of license plates.

Warning Signs

Signs warning motorists that red light cameras are being used are typically required by law or ordinance but, whether required or not, should be posted as part of the driver awareness and education process. These warning signs may be placed in the following locations at photo-enforced intersections:

- In advance of photo-enforced intersections.
- At photo-enforced intersections, typically on the far side traffic signal pole.
- On all approaches into an area where red light camera systems are used for red light running.

Warning signs placed on all approaches into an area, while used to satisfy legal requirements in some jurisdictions, are appropriate as supplemental warning signs but not as the primary warning for motorists. Advance warning signs should be installed at photo-enforced intersections.

All advance warning signs should be clearly visible and compliant with the MUTCD [17].

Traffic Signal Yellow Times

The MUTCD and ITE recommended practice on the length of yellow interval times provides adequate and proper direction to practitioners. Yellow times should be established in accordance with the MUTCD [17] guidelines and the ITE [9] informational report for methods for calculating yellow time intervals.

Changes in the yellow times after red light camera systems are in place and operational will affect the number of photographed violations, increasing the number of violations when yellow times are shortened and reducing the number of violations when yellow times are lengthened. Where changes in the yellow times at intersections with red light camera systems are required as the result of updated speed surveys or other factors, the changes should be clearly described in public information announcements. Providing warning notices for a reasonable amount of time after the change is particularly important for violations recorded at intersections where the yellow interval has been shortened.

System Selection and Technologies

The most widely used red light camera systems employ film-based cameras and inductive loop vehicle detection technologies. However, other red light camera technologies have become available over the past five years, most notably technologies that employ digital camera equipment where photographic data, including streamed video clips, may be immediately downloaded for processing using leased telephone line or microwave communications. Additionally, red light camera systems that use video-based and radar vehicle detection methods, as well as systems that employ overhead camera placements and floodlighting equipment as an alternative to the curb-based placements, are used by many State and local agencies.

A red light camera system consists of the following on-the-street components:

- Camera Units.
- Intersection Lighting.
- Camera Housing and Supporting Structure.
- Vehicle Detection.
- Communications.
- Warning Signs.

Each of these components is reviewed in the following sections.

Camera Unit

There are three general types of cameras units used to automatically record red light violations. The types of camera units used in red light camera systems include:

- 35mm Conventional Film Units.
- Digital Still Picture Units.
- Digital Video Units.

Each type has both pros and cons, as shown in table 5.

Table 5. Camera Units Compared

Camera Unit	Pros	Cons
35 mm	Best resolution	Collection and development of film
Digital Still	Digital format Ease of use No film collection or development.	Needs communication links between cameras and processing center Comparatively poor resolution
Digital Video	Provide video clips of alleged violations Provides circumstances in which violations occur	Impression of surveillance Needs communication links between cameras and processing center

Intersection Lighting

Additional intersection lighting is required in conjunction with the operation of the camera units. The additional lighting will need to be installed in accordance with the equipment manufacturer's specifications, as well as with State or local ordinances that govern the amount of lighting that is permitted in the driver's field of view.

For camera units that record violations with one or two photographs or digital images, flash units synchronized with the camera shutter provide additional lighting at the intersection at time of exposure so vehicle license plate and drivers, if local or State law allow, can be more clearly photographed. Typically, one flash unit is installed as an integral part of the camera housing. Additional flash units may be installed at intersections where there are more than two lanes being monitored or to maximize the amount of backlighting in the vehicle interior as it traverses the intersection.

For camera units that record a video clip for each violation, continuous additional lighting will be considered. This may be provided by curb or overhead mounted lighting equipment, as specified by the equipment manufacturer.

Camera Housing and Supporting Structure

The types of camera housing and supporting structures will depend on the type of red light camera system being installed.

Curb-mounted red light camera systems, the most common type currently being employed by State and local agencies, need a camera housing enclosure that is mounted on a pole. The camera unit housing should be weather and damage resistant, and contain a locking mechanism to protect the system from vandalism. Additional poles may be employed for auxiliary flash lighting units. For digital camera systems, a separate enclosure for the data storage and communications equipment is also required at the intersection.

The poles for curb-mounted red light camera systems should be tall enough to provide the necessary angle of view to clearly record violations at the intersection. There are at least two types of poles currently in use. The first, a hinged pole, lowers the camera housing on a hinge located in the center of the pole. A second type, a solid pole, utilizes a motorized "elevator" to raise and lower the camera housing.

Overhead-mounted red light camera systems normally require curb-mounted poles with cantilever arms extending over the traffic lanes. Camera and flash units are mounted on the cantilever arms as required for system operation. Red light camera systems of this type provide an increased field of view that is especially advantageous for red light camera systems on wider arterial streets as well as enhanced lighting for enhanced photographic data quality.

Some jurisdictions have found that they can afford only a limited number of red light camera systems. By installing red light camera housings at problem intersections, and periodically moving the actual cameras from housing to housing, gives motorists the impression that cameras are omnipresent and reduces red light violations throughout the community.

Vehicle Detection

Vehicle detectors are used to trigger the camera to record a vehicle running a red light. Different vehicle detection technologies are available for this purpose.

Most red light camera systems employ pairs of inductive loops installed near the intersection at a location suitable for showing that a violation has occurred. It is critical for the system design and operation that the inductive loops be installed in the appropriate locations, consistent with the agency's definition of a violation. Red light camera systems may also employ piezoelements, video-based equipment, or radar devices for vehicle detection and tracking, as an alternative to, or in conjunction with, inductive loop detectors.

The placement of the vehicle detectors is critical to the integrity of the red light camera system and the citations developed from the photographic data.

For red light camera systems that document violations with two photographs, the first photograph should be taken to show the motor vehicle that will be running the red light, at a location immediately before it enters the intersection against a red traffic signal indication. The vehicle detection equipment should be configured to detect the presence of the vehicle at the desired location and to initiate the first photograph being taken with the vehicle at that location. If the vehicle is detected after it has already entered the intersection, it cannot be determined with certainty from the photographs that the vehicle entered the intersection illegally and consequently, a citation should not be issued. The second photograph is taken after the vehicle has entered the intersection, at a time interval after the first photograph calculated to provide the best view of the vehicle and its license plate, and where required, the driver's face.

For red light camera systems that document violations with video clips that show the vehicle running the red light continuously starting at a location before the vehicle enters the intersection against the red traffic signal indication, vehicle detection should be configured so that the video clip recording is initiated at an appropriate location.

The placement of inductive loop detectors immediately in advance of the intersection stop line for vehicle detection may require that existing stop line loop detectors used for the traffic signal operations need to be abandoned, relocated, or replaced with another type of vehicle detection system, such as video-based detection. Generally, a solution that accommodates vehicle detection requirements for both traffic signal operations and the red light camera system can be developed although there may be some additional costs for vehicle detection associated with the installation of the red light camera system equipment under these circumstances.

Communications

For digital camera units, a communications link with adequate bandwidth should be provided from the intersection to a location where the violations data is processed. The required communications may be implemented using

State and local agency fiber optics, leased high-capacity telephone lines, or microwave technologies.

No communications outside of the intersection are required for 35mm conventional film camera units.

Communications links are normally required to support certain functions related to citation data processing, including access to vehicle registration and driver's license databases, data transfers to and from traffic court data processing systems, and on-line inquiries or payments from persons receiving citations.

Warning Signs

Refer to earlier section for guidance on warning signs.

ENGINEERING DESIGN OF RED LIGHT CAMERA SYSTEMS

The red light camera system installation plans should be prepared and signed by an appropriately licensed engineer. Installation plans should be prepared in accordance with the system manufacturer's standard plans and technical specifications, and with State and local agency standard plans and specifications for public works and traffic engineering improvements. The plans should address the placement of the red light camera system equipment and related components, including:

- Camera equipment.
- Camera housing and supporting structure.
- Intersection lighting.
- Vehicle detection system.
- Communications.
- Pullboxes, conduit runs, and conductor schedule.
- Electrical service.
- Warning signs.

There are currently no standard plans and specifications for the acquisition and installation of red light camera systems, except for the plans and specifications provided by the manufacturers and standard plans and specifications that have been developed by State and local agencies for their own use and application. The U.S. Department of Transportation (USDOT), through a cooperative agreement with the International Association of Chiefs of Police (IACP), are developing red light camera systems performance specifications and testing laboratories to ensure the accuracy and reliability of these systems.

The installation plans should be processed through the appropriate State or local agency plan review and permitting procedures.

Red Light Camera System Installation

Where a contractor does the installation work, the normal construction inspection procedures employed by the State or local agency should be carried out for the installation of the red light camera equipment. Proper installation includes:

- Installation consistent with the equipment manufacturer's guidelines and State or local agency specifications.
- Inspection of all installation work by State and local agency officials and, where necessary, by the project engineer.
- Testing of the red light camera equipment prior to its cutover for unattended operation.
- The preparation of as-built drawings that reflect actual construction conditions.

Installations should be thoroughly inspected before testing begins. A comprehensive testing program should then be conducted using both simulated and actual traffic before the system is placed into unattended operation. No warning letters or citations should be issued until it is determined that the system is working accurately and reliably.

Operation and Maintenance

As with any integrated system, every element of a red light camera system should function properly for the system to produce the desired results. In addition to proper design and installation, procedures to ensure the proper operation and maintenance of the system should be developed and implemented by the State and local agency.

Proper operation should be consistent with the manufacturer's instructions and the documented operational procedures that have been developed, reviewed, and approved by all parties involved. Periodic checks and audits to verify that it continues to operate properly should also be conducted.

Proper maintenance should include both preventive and corrective maintenance. Preventive maintenance should be performed on a regular basis. Tests of operational performance should be conducted regularly, and actual operational results examined constantly in order to identify any variation from specified performance. If any flaw in the system operation or performance is detected, the issuance of citations should be immediately stopped and any citations previously issued with the possibility of flawed operation or performance should be withdrawn.

Red light camera system operations and maintenance should include the following tasks and functions:

- Collect images of recorded violations and related violations data from photo-enforced intersections.
- Inspect camera and vehicle detection system operations.
- Perform preventative maintenance and cleaning.
- Identify defective equipment and make repairs or replace the equipment.
- Store recorded violations data.
- Review recorded violations data to identify violations.
- Identify vehicle registered owner.
- Prepare draft citations for review and approval.
- Prepare and mail citations to vehicle registered owners.
- Answer telephone inquiries.
- Schedule violator appointments.
- Process vehicle registered owner certifications regarding driver identity at the time of the violation.
- Provide court-requested information and support court hearings.
- Prepare monthly progress reports.

Citation Data Processing

The procedures and methods employed for system operations should be designed to ensure the preservation of the chain of custody of evidence for each recorded violation so that backup data and documentation can be easily retrieved when needed. The procedures and methods used for system operations should be comprehensive, clearly documented in writing, and followed without exception.

Citation data processing should be carried out in a secured facility using a data processing system with appropriate security features and firewalls. All personnel, especially those with access to motor vehicle registration and driver's license databases, should be cleared with appropriate background checks.

Internal quality control is essential and should be achieved by the use of two separate internal reviews of each violation, periodic audits by independent law enforcement or engineering staff, and other procedures. Procedures, especially important to ensure quality control, should be developed for each of the following areas:

- Guidelines to be applied for issuing a citation. In other words, a very specific definition is needed to identify what constitutes a red light running violation.
- Citation review and approval requirements, including provisions for the procedure to be used when the time to review is shortened, traffic officers are not available to conduct the reviews, or the number of citations is larger than usual.
- Quality assurance audits, to be conducted by trained traffic officers for randomly selected sample of recorded violations on a periodic basis.

Only a qualified law enforcement officer should be authorized to issue a citation. Citations should not be created prior to review of appropriate evidentiary material by the officer. Under no circumstances should a citation be issued when the officer expresses any lack of confidence that a properly documented and provable violation has occurred.

System Maintenance

Periodic inspections and preventative maintenance should be required to ensure that the equipment is functioning properly. Service and inspection logs should be maintained to document the inspections and preventative maintenance activities. The service and inspection logs may be required at court hearings to confirm that the red light camera equipment was functioning properly at the time that the violations were recorded.

As part of the periodic preventative maintenance, the camera unit and housing should be thoroughly cleaned and the camera unit activated in its "test" mode and confirmed to be operating correctly. The condition of the camera housing and mounting structure, camera unit, vehicle detection system, and warning signs should be inspected and the conditions noted in the service and inspection logs.

On-Going System Assessment

Continual analysis of violation and crash data, with community input, is an important element of a successful red light camera program. Adequate funding should be provided to assure the necessary data analysis, problem identification, and problem diagnostic review work tasks are undertaken.

Red light running camera enforcement efforts should be monitored, with adequate pre- and post- installation study periods, in order to measure the program's effectiveness. Timely collection and reporting of crash data is an important part of the monitoring process, as are control sites with no photo enforcement so the effects of camera enforcement can be distinguished from other external effects.

The steering committee should meet on a regular basis. Regular agenda items should be to review the data of violations and citations issued with a discussion of any changes or trends noted. Input from the State or local agency's traffic engineering department and street maintenance department should include regular updates on planned traffic signal modifications or street improvements construction that could impact the operation of the system. Discussion should be encouraged on whether program objectives are being met through the deployment of red light camera systems or whether alternative measures should be applied. The group should have input to the regular prioritization of intersections targeted for safety-related improvements.

A monitoring program based on the timely collection and reporting of crash data is needed. These crash data should include control sites with no photo enforcement so that the effects of camera enforcement can be distinguished from other external effects. Responsibilities for the collection and reporting of crash data need to be established and clearly defined. Traffic safety professionals need to review intersection safety issues and conduct diagnostic reviews of intersections identified from the crash data tabulations as warranting safety-related improvements.

Regular reports on the public awareness and information campaign should be prepared and reviewed. Public use of the web site and telephone information systems should be monitored.

ON-GOING PUBLIC INFORMATION AND EDUCATION

An on-going public information and education campaign is needed to assure the motoring public that the red light running camera program is being operated in the most effective, efficient, and fair manner possible. Public information and education efforts begin before installation, but do not end when the system is fully operational.

The on-going educational program should be designed to combat red light running, in general, as well as to provide information related to the operation of the red light camera equipment. Where possible, the on-going public information and education program should be developed and delivered in a way so as to address any specific populations or conditions that have been identified as contributing extensively to the red light running problem.

The on-going public information and education program should use various media, including the print and broadcast media, to communicate the problem, the program and the results. The agency should monitor the effectiveness of the educational program in order to achieve maximum effectiveness and public support for the red light camera program.

APPENDIX A. PHOTO RED LIGHT ENFORCEMENT LEGAL CONSIDERATIONS

NOTICE

The National Highway Traffic Safety Administration (NHTSA) and the Federal Highway Administration (FHWA) has compiled and distributed this information as a legal guide only. This material is not intended to be a complete treatment of every jurisdiction's laws and court decisions related to photo red light enforcement. Instead, this material includes highlights and examples of court decisions, and discusses issues that users engaged in photo red light enforcement should consider.

Due to the dynamic nature of law enforcement and the evolution of technology, it is important that each department review this information to verify that it is consistent with applicable, current State and local law and regulations, and with department policy and procedure. *This information is NOT intended to substitute for the advice of legal counsel.* You should speak with your legal advisor, and/or local prosecutor, about the sufficiency of your department's manual, policy, curriculum, and training program on this subject. This material should not be used as the sole basis for compliance with any law or regulation, and *departments should NOT rely on this material as a legal defense* in any civil or criminal action. Remember that new court decisions and amendments to the law could change the material in this appendix.

Photo red light enforcement is a relatively new law enforcement tool. Thus, case law is not well established. Although the few cases involving photo red light raised constitutional issues, the decisions were based upon procedural grounds, never answering the ultimate question – is it constitutional? The ruling on the Motion to Dismiss citations issued under San Diego, California's photo red light program (under appeal as of the preparation of this report), found the program constitutional. However, this ruling is not binding and only provides insight into the court's reasoning.

Automated speed enforcement, a relatively new enforcement tool as well, shares common legal issues with photo red light enforcement – such as the registered owner presumption, notice, procedural, constitutional issues, etc. Most automated speed cases have also tended to avoid constitutional questions. Some issues (e.g., chain of custody, service of process issues, registered owner presumption) have been addressed, but these decisions tend to be highly fact-dependent and/or are based on State statutes.

Many questions remain. The answer to these questions may be gleaned from cases not specific to automated enforcement. Existing case precedent dealing with evidentiary issues of older enforcement techniques will shape the use of automated enforcement evidence in the future. Law enforcement will use the same criminal procedures as are applicable to the collection (search and seizure), preservation (chain-of-custody), and discovery of other types of evidence.

It is most important to note that although the courts will borrow from established case law to determine case law regarding automated enforcement, the path will most likely be contorted. The law is known for nuances. Thus, subtle distinctions between photo red light programs may affect a court's decision and produce seeming inconsistencies.

Most importantly, the classification of the photo red light violation, as either a civil or criminal violation, will dramatically effect decisions. Similarly, as in the San Diego photo red light program, the enabling statute may impact the admissibility of the evidence (see page 51 for enabling statutes). A State's surrounding body of law and the manner in which the program is conducted will also impact the viability of the photo red light program and the success or failure of challenges to the program. Without assessing merit, the following are some of the procedural and substantive issues that may be generated by photo red light enforcement.

Procedural Issues:

- Authentication of photographs
- Chain of evidence of photographs
- Compliance with enabling statutes
- Foundation: Device reliability (maintenance, checks for accuracy, training of personnel involved in the process)
- Misuse or dissemination of photographs
- Municipal drafting
- Notice – compliance with applicable state rules for service
- Proper notice of use of photo red light enforcement (signs)
- Standing – who can bring an action, when, and where

Substantive Issues:

- Administration of the program violates Fourteenth Amendment Due Process rights
- Confrontation rights (6th Amendment right)
- Equal Protection (disparate treatment for public, police, rental, corporate, out-of-state vehicles, motorists cited by police)
- Fifth Amendment right to remain silent (for statutes requiring affidavit as to who was driving)
- Mailing a citation that requires appearance is a seizure subject to the Fourth Amendment
- Photographing a motorist is a search subject to the Fourth Amendment

- Pre-charging delay (delay between the violation's occurrence and receipt of notice) – Fourteenth Amendment Due Process
- Presumption that the registered owner is the driver impermissibly shifts the burden of proof
- Privacy – violation of State privacy laws
- Revenue generation: selection criteria for photo red light enforcement, light phase timing
- Substantive Due Process - Privacy

The above are all issues that are likely to continue to be the subject of legal review and refinement. Monitoring their long-term clarification through legal proceedings is to be advised for all jurisdictions adopting red light camera enforcement systems. The remainder of Appendix A reviews current relevant case law examples and in doing so illuminates the types of issues that have been raised.

PHOTO RED LIGHT CASE LAW SYNOPSIS

Dajani v. Governor of Md., No. CCB-00-713, 2001 U.S. Dist. LEXIS 982 (D. Md. Jan. 24, 2001) (unreported).

Facts: The defendant was charged with a photo red light violation and convicted. In this jurisdiction, photo red light violations are civil and not considered moving violations. Insurance companies may not consider the convictions.

Issue: The defendant appealed to the Federal district court, requesting the court declare the statute unconstitutional. The defendant alleged the photo red light statute violated the Sixth Amendment's Confrontation Clause and the Fourteenth Amendment's Due Process Clause.

The court upheld the conviction on procedural matters (lack of Federal jurisdiction and lack of standing) without comment on the constitutional issues. The Fourth Circuit Court of Appeals affirmed the District Court's decision. (Dajani v. Governor of Md., No. 01-1179, 2001 U.S. App. LEXIS 17303 (4th Cir. 2001).

Kovach v. District of Columbia, 805 A.2d 957 (D.C. 2002)

Facts: The defendant paid, without contesting, a photo red light citation. Subsequently, the police department "decided to remove the camera because it was observing an inordinate number of people running the light, which was confusing to motorists." *Id.* at 959. Outstanding fines were dismissed, but those motorists who had paid were not reimbursed.

Issue: The defendant appealed, alleging the District's decision to forgive some, but not all, violations violated the Fifth and Fourteenth Amendments.

The court upheld the conviction because "in failing to contest the infraction, appellant effectively acknowledged liability for running the red light." The court also rejected the defendant's argument that the confusing placement of the stoplight created "manifest injustice." The defendant "has no standing to challenge the decision unless . . . he was confused . . ." *Id.* at 962-63.

Structural Components Int., Inc. v. City of Charlotte, No. C0A102-200 (N.C. Ct. App., Nov. 19, 2002) (unreported – not final until expiration of rehearing period)

Facts: The president of Structural Components received a photo red light citation for one of its vehicles. In this jurisdiction, violations are civil. Structural Components contested the violation at a "review hearing."

Issue: Upon conviction, Structural Components (plaintiff) filed suit in the superior court alleging negligence (by failing to establish reasonable guideline, failure to govern the program in a reasonable manner, and failure to provide a reasonable appeals process) and civil rights violations (State/Federal due process and equal protection).

Upon defendant's (the City and Lockheed Martin) motion to dismiss, the court determined it lacked jurisdiction and dismissed the action. Structural Components appealed. The appellate court affirmed the trial court's dismissal on procedural grounds (waiver of the negligence action for failure to properly state issue in appeals brief and, because one cannot recover monetary damage for a procedural due process violation involving a civil penalty, failure to state a claim). The court noted the proper avenue to challenge the constitutionality of the statute was by certiorari to the superior court (which Structural Components had not used) and the present statutory scheme provided an adequate method for challenging the legality of the program.

City of Commerce City v. Colorado, 40 P.3d 1273 (Colo. 2002)

Issue: Commerce City challenged whether the Colorado statute

(COLO. REV. STAT. § 42-4- 110.5 (2002)), which authorized the photo red light program, infringed upon the City's "home-rule" powers. Noting that the program involved a "mix" of state and local concerns and, where conflicts arose, State concerns prevail, the court affirmed the validity of the program.

People v. John Allen (In re Red Light Camera Cases), No. 57927SD (Cal. Super. Ct. Aug. 2001) (order denying motion to dismiss) (available at http://freedom.gov/auto/ cases/ sdmotion.asp). *This case remains under appeal. This order is presented to illustrate issues that may arise with photo red light enforcement.*

Facts: Defendants in a photo red light case filed a motion to dismiss alleging failure to comply with the authorizing statute (section 21455.5 of the California Vehicle Code). In this jurisdiction, the violation is criminal and a conviction is entered onto the driver's license record.

Issue #1: The defendants contended the photo red light program was not operated by a government agency in cooperation with a law enforcement agency as required by the authorizing statute.

The court noted "once the construction process was begun, there was very little City involvement." The City did not inspect the project when complete and the "entire process of installation and calibration of the camera equipment, putting film into the cameras, unloading the cameras, developing the film, maintaining the camera equipment, and reviewing the photographs to make the initial determination as to whether or not there was a violation and whether the alleged violator can be identified, is done by Lockheed Martin. Further, once Lockheed determines that a citation will not [be] issue[d], that decision is not reviewed by the City.

If Lockheed decides a citation should [be] issue[d], it reviews Department of Motor Vehicles' information . . . prints the citation, including printing the signature of the sergeant in charge of the program on the citation. The first time the City becomes involved is when the police department receives the citation which has already been printed." The police review copies of the photographs and the digital information to determine whether the citation should be issued. If a citation is issued, Lockheed mails it . . ." The court found the City had "no involvement with, nor supervision over, with the ongoing operation of the system" and "[t]he Legislature did not contemplate such a lack of participation by the City" when it authorized a government agency to "operate an automated enforcement system." Thus, the program violated the statute.

Issue #2: The defendants contended the signs were inadequate.

The statute required signs "clearly indicating the system's presence, visible to traffic approaching from all directions." The posted signs were 24" by 30." Based upon testimony of police officers as to the signs visibility and the lack of evidence drivers were not able to see the signs, the court found the signs adequate.

Issue #3: A related statute (section 40520 of the California Vehicle Code) required photo red light violations to be accompanied by an affidavit of non-liability, information as to what constitutes non-liability, information as to the effect of executing the affidavit, and instructions for return. The defendants alleged this procedure was "unconstitutional because it requires innocent people to testify against each other."

The court noted the section was a legislative attempt to prevent blanket immunity for corporate and rental agencies vehicles and provides a method for the registered owner who is not driving to avoid liability. Without elaboration, the court determined the statute compliant with due process and "a legitimate exercise of the police power in an attempt to issue citations to the actual driver who violated the red light."

Issue #4: The California Penal Code (section 959.1) requires pleadings (citations) be sworn before an officer entitled to administer oaths. The defendants alleged that "no officer swears to the facts because the signature is affixed electronically before it is sent to the police and the officer who reviews the citation is not the sergeant whose signature appears on the citation." The reviewing officer merely stamps his ID number below the signature.

The court noted that pleading defects (i.e., minor errors in the pleading document) that do not prejudice a substantial right do not justify dismissal.

Issue #5: The defendants argued that because the City did not comply with statutory provisions regarding the "operation' of the program, all citations must be dismissed. In making its determination the court looked at the following issues.

Issue #5A: Was the delegation of authority constitutional?

Although, the City had delegated the tasks of evidence collection and determining who will not be cited to Lockheed Martin, the police retained the "ultimate authority to determine who will be prosecuted." Thus, the delegation was not unconstitutional.

Issue #5B: Is the fee paid to Lockheed Martin a contingency fee and if so, what is the legal effect?

Because Lockeed's payment was contingent upon a conviction, the fee was deemed a contingency fee.

The court indicated that Lockheed was "supposed to be a neutral evaluator of the evidence" and "should not have a financial interest in the outcome." The court reasoned that because the statute mandated a government agency "operate" the program, the purpose was to guarantee, "information obtained from the red light cameras would be trustworthy. The potential conflict created by a contingent method of compensation further undermines the trustworthiness of the evidence which is used to prosecute the red light violations."

Issue #5C: Does the delegation, without statutory authority, which operates on a contingent fee basis violate due process such that it requires a dismissal of pending actions?

The court noted that the threshold question in a due process challenge to executive action is whether the behavior is "so egregious, so outrageous, that it may fairly be said to shock the contemporary conscience." In this case, the court held the conduct did not rise to that level.

Issue #5D: Is the photo red light evidence admissible?

The court indicated that "where evidence is obtained from sources subject to legislative standards, there should be substantial compliance." The court noted that "there is no authority in the Vehicle Code for unsupervised private operation of a red light camera system. Therefore, there is not substantial compliance with the safeguards required by the statute. Such a lack of authority, combined with the collection based compensation, result in evidence lacking foundation. Without foundation, the evidence is not relevant and is not admissible."

Accordingly, the court did not grant the motion to dismiss, but rather granted a motion to exclude the evidence.

Office of the Attorney General of the State of Texas, Opinion No. JC-0460, 2002 Tex. Op. Atty. Gen. 20 (2002).

Issue: Could a city pass an ordinance authorizing a photo red light program and could violations be civil, rather than criminal?

Based on Texas law (which deemed red light violations criminal), the Attorney General opined a city could authorize a photo red light program to identify violators, but could not make violations civil.

Office of the Attorney General of the State of Tennessee, Opinion No. 01-004, 2001 Tenn. AG LEXIS 6 (2001) (available at http://www.attor neygeneral. state.tn.us/op/ 2001/OP/OP4.pdf).

Issue: The Attorney General's Office was tasked with determining whether, pursuant to inherent police power, a city had authority to enact ordinances allowing photo-enforcement.

Without addressing specific constitutional issues, the Attorney General's opinion concluded that the use of photo-enforcement did not conflict with any State statute. In a footnote, the opinion noted photo-enforcement has "generally been viewed as a permissible exercise of State and local government police power which is not violative of Federal or State constitutional provisions."

Office of the Attorney General of the State of Nebraska, Opinion No. 00001, 2000 Neb. AG LEXIS 1 (2000) (Available at http!!:www.ago.state. ne.us!opinion! index.html).

The Attorney General's office was tasked with assessing the constitutionality of proposed legislation involving photo red light enforcement. The Attorney General offered the following opinions:

Issue #1: Procedural Due Process

The proposed legislation permitted a defendant to contest the violation in a county court and assumed that proper notice would be provided. Thus, the Attorney General opined the proposed legislation would comply with the procedural due process requirements of reasonable notice and an opportunity to be heard.

Issue #2: Substantive Due Process

Substantive due process guarantees individuals protection from arbitrary government action. The Attorney General noted that due process is satisfied if the government has the power to act on the subject matter, if they did not act capriciously or in a discriminatory manner, and if there was a reasonable relationship to a proper governmental purpose.

The Attorney General opined that the proposed legislation complied with substantive due process because protecting public safety is a proper subject matter and the legislation was rationally related to that interest.

As to the registered owner presumption, the Attorney General opined this was also a "proper exercise of the State's police power" similar to holding the registered owner of a parked vehicle liable.

Issue #3: Equal Protection

The Attorney General noted the similarities of the Nebraska and U.S. Constitution in that equal protection challenges not involving a suspect class or fundamental right are tested only for rationality. A Nebraska Supreme Court decision (State v. Michalski, 221 Neb. 380, 377 N.W.2d 510 (1985)) had held that driving is not a fundamental right, and that drivers were not a suspect class.

The Attorney General opined that the classification would be between two types of drivers: (1) those individuals cited directly by an officer who receive a criminal penalty, and have the conviction recorded on their driver's license; and (2) those individuals cited by the photo red light program who are subjected only to civil penalties and no recordation on their driver's license.

The Attorney General noted that, although the purpose of the legislation was not set forth, the apparent purpose was to reduce the hazards of running red lights. Thus, the Attorney General concluded that, given the "wide latitude" and deference to the legislative process, the legislation met the rational basis standard and the proposed law would comply with Equal Protection rights.

Oregon v. Dahl, 57 P.3d 965 (Or. Ct. App. 2002).

Facts: An officer operating a photo radar unit photographed the defendant's vehicle exceeding the posted speed limit. The defendant was the only registered owner. The officer observed the violation, but did not effect an enforcement stop and could not identify the driver. At trial, a witness commented that the defendant failed to provide a sworn certificate of innocence as permitted by statute.

Issue #1: The defendant contended the Oregon statute which establishes a presumption that the registered owner of a vehicle is the driver impermissibly shifts the burden of persuasion.

An Oregon statute (OR. REV. STAT § 153.030.1) provides that unless excepted, criminal procedure laws apply to traffic violations. However, a different statute (OR. REV. STAT § 153.076.2) provides that traffic violations must be proved by a preponderance of the evidence (a civil standard). Because this statute authorized a civil standard of proof, the court reasoned a civil standard also applied to the presumption. Therefore, the burden shift was permissible.

Issue #2: The defendant contended that, even if the violation is civil, the Oregon presumption statute violated due process standards.

The court noted that both U.S. Supreme Court (Bandini Petroleum Co. v. Superior Ct., 284 U.S. 8 (1931)) and Oregon State court decisions required a "rational connection" between the fact proved and the ultimate fact presumed. The defendant argued that "vehicles usually have more than one key, licensed drivers outnumber registered vehicles, and vehicles commonly are borrowed or stolen, all of which indicate that vehicle are often driven by someone other than their owner." The court, although acknowledging that vehicles are often driven by non-owners, found that "it is not irrational for the legislature to presume that vehicles are often driven by owners" and "we need not decide what facts are more likely to be true; the rational connection test does not require adoption of the best or most persuasive explanation." Thus, the Oregon statute did not violate due process. *Id.* at 968-969.

Issue #3: The defendant contended a witness reference to her failure to submit a sworn certificate of innocence violated her statutory and constitutional right to remain silent.

The Fifth Amendment provides that no person "shall be compelled *in any criminal case* to be a witness against himself." The court indicated that the defendant had failed to identify how she could have been exposed to any criminal responsibility. Thus, "her constitutional right was not implicated." *Id.* at 969.

Section 810.439 provides a defendant in a traffic violation case an opportunity to avoid trial by submitting a certificate of innocence. The defendant may disregard that opportunity. The court "assumed without deciding" that the witness's comment impermissibly infringed on the defendant's statutory right, however, the court also stated "there was no indication that the trial court relied on that testimony in making its decision." Thus, the court found the defendant was not prejudiced by the comment. *Id.*

McNeil v. Town of Paradise Valley, No. 01-17003, 2002 U.S. App. LEXIS 17306 (9th Cir. Aug. 12, 2002). Not Published – Check with Court Rules. The case is presented to illustrate issues that may arise with photo red light enforcement.

Issues: McNeil appealed the district court's dismissal of alleged civil rights and Racketeer Influenced and Corrupt Organizations Act (RICO) violations premised on the issuance of an automated speed citation. The facts and basis for these contentions was not clearly set forth.

However, it appears that McNeil contended the mailing of a traffic citation to the registered owner was a seizure and the process was in violation of due process.

Without elaboration, the court found municipalities cannot constitute a RICO enterprise. Further the court indicated that, because a seizure requires intentional physical control, the mailing of a citation is not a seizure. As for the due process claim, the court indicated that the challenge to the citation in municipal court was sufficient.

Oregon v. Clay, 29 P.3d 1101 (Or. 2001).

Facts: An officer operating a photo radar unit photographed the defendant's vehicle speeding. The officer did not effect an enforcement stop and did not know the identity of the driver in the radar photo. Subsequently, a citation was issued and mailed to the defendant. The defendant did not appear at trial, but rather was represented by counsel. No evidence was presented on behalf of the defendant. The State presented no direct evidence that the defendant was the registered owner, but rather relied on witness testimony and an "official duty" presumption to establish the defendant as the registered owner. Upon being found guilty, the defendant appealed, contending the State had failed to prove that she was the registered owner of the vehicle. The Oregon Court of Appeals upheld the conviction and the defendant appealed to the state supreme court.

In this jurisdiction, the registered owner is presumed to be the driver – see Oregon Law 1995, Chapter 579, sections 1-3 later codified to Oregon Revised Statutes §§ 810.438-810.439. Oregon statute section 811.123 requires proof that a particular person was speeding.

Issue: The defendant contended there was insufficient evidence to permit the trier of fact to find that she was the registered owner of the vehicle.

The court indicated that it did not "perceive any evidentiary basis . . . that would permit a trier of fact to find that the defendant was the registered owner of the speeding car." *Id.* at 1103. The percipient witness could not identify the driver and there was no evidence to conclude the defendant was the registered owner (which would have invoked the presumption that the registered owner was the driver).

The court indicated that because an officer had the authority, not a duty, to send the citation, the presumption that an "official duty had been performed" was not applicable.

Because they could not prove the notice had been mailed to the registered owner, they could not prove the defendant was the registered owner. Because they could not prove that the defendant was the registered owner, the presumption that the registered owner was the driver was not applicable.

Oregon v. Weber, 19 P.3d 378 (Or. Ct. App. 2001).

Facts: An officer operating a photo radar unit observed the defendant's vehicle speed. The unit photographed the vehicle. Subsequently, the defendant was mailed a citation.

Issue #1: The defendant contended the inscription (indicating vehicle speed) on the photograph was impermissible hearsay.

The court indicated that, by statutory definition (Oregon Evidence Code 801), hearsay is a statement by a declarant and a declarant is a person who makes a statement. A machine, not a person, made the inscription on the photograph. Thus, the hearsay rule is inapplicable.

Issue #2: The defendant contended the court should have excluded the photograph on chainof-custody grounds because the state offered no evidence as to "who picked up the film from the station, what happened to the film, how it was handled, or what was done to it prior to the citation and photograph being returned to the police station six days later."

The court indicated that, "given the totality of circumstances, the trial court was well within its discretion in determining that there was no appreciable likelihood of alteration or tampering and that no further foundation was required." *Id.* at 381-82.

Issue #3: The defendant contended the automated speed enforcement unconstitutionally shifts the burden of proof of the offender identity.

The court ruled the defendant had failed to use the proper judicial procedure to preserve this issue.

Issue #4: The defendant contended the time delay (between the occurrence of the violation and the mailing of the notice) violated her Fourteenth Amendment due process rights.

The court indicated that "for a precharging delay to give rise to a due process violation, a defendant must show both substantial prejudice to his right to a fair trial and that the delay was done intentionally to gain a tactical advantage." The court found the defendant failed to establish the state intentionally delayed the notice to gain a tactical advantage. *Id.* at 385.

Bentley v. West Valley City, 21 P.3d 210 (Utah 2001).

Issue: Plaintiffs, who received automated speed enforcement citations, sought reimbursement of fines alleging the automated enforcement program violated Utah Code section 41-6-52.5. None of the plaintiffs had challenged the program during the criminal proceedings. The court ruled on procedural grounds finding the plaintiffs failed to assert an "actionable civil theory under which criminal fines are recoverable . . . ."

Anchorage v. Baxley, 946 P.2d 894 (Alaska Ct. App. 1997).

Facts: The defendants received automated speed enforcement citations. At trial, numerous witnesses testified to the reliability of the speed enforcement device. However, the trial court found the witnesses' financial interest in the acceptance of speed enforcement units tainted their credibility. The magistrates found that, absent independent corroboration as to the reliability of the device, results were not admissible.

Issue: The city appealed seeking a ruling that automated speed enforcement evidence was admissible without corroboration.

The court indicated the case was moot because "we would only review the magistrates' decision to determine whether the evidence presented would allow a reasonable fact finder to conclude that the municipality had failed to prove its case." And, given the magistrates' dim view of the witnesses' credibility, no reversible error occurred. *Id.* at 598-99.

West Valley City v. McDonald, 948 P.2d 371 (Utah Ct. App. 1997).

Facts: The defendant received an automated speed enforcement citation and requested a jury trial. The state amended the complaint to a lesser charge (which did not warrant a jury trial). Subsequently, the defendant was convicted.

Issue: The defendant appealed claiming that reducing the charge deprived her of her statutory right to a jury trial.

The court upheld the conviction

Tonner v. Paradise Valley Magistrate's Court, 831 P.2d 448 (Ariz. Ct. App. 1992).

Facts: An automated speed enforcement citation was mailed to General Motors Acceptance Corporation (GMAC), the registered owner. GMAC forwarded the notice to Tonner and mailed a copy of its transmittal letter to the court.

The court reissued the notice to Tonner. Tonner failed to reply or appear. The court entered an order for a civil sanction (fine).

Issue: Tonner filed an action to vacate the sanction arguing lack of personal jurisdiction based upon improper service of notice.

The court indicated that under Arizona civil procedure rules (ARIZ. R. CIV. P. 4.1c), service is not complete unless acknowledged. As Tonner failed to reply, service was not complete. Without service, the court lacked jurisdiction by which to sanction Tonner.

Office of the Attorney General of the State of South Carolina, (No Opinion No.) 2002 S.C. AG LEXIS 209 (2002).

The Attorney General re-evaluated[1] the use of automated traffic enforcement and concluded that "general case law and other authority reviewed herein support the conclusion that a properly drafted statute authorizing use of photo-radar or similar forms of automated traffic enforcement would pass constitutional muster. These authorities have reviewed automated traffic enforcement from a variety of constitutional perspective include the Due Process and Equal Protection Clauses, the Fourth Amendment's protection against unreasonable searches and seizures, the Sixth Amendment's right to present an adequate defense, as well as the federal and state constitution's right to privacy. The general consensus is that automated traffic enforcement is constitutional."

"Of course, the constitutionality of any statute authorizing automated traffic enforcement would depend, in part, upon a well drafted statute."

See also:

1) Office of the Attorney General of the State of Mississippi, Opinion No 2000-0068, 2000 Miss. AG LEXIS 113 (2000) indicated that, prior to implementing a photo red light program, a municipality would need statutory authority allowing citation of the registered owner of a violator's vehicle.
2) Office of the Attorney General of the State of South Carolina, 1996 S.C. AG LEXIS 54 (1996) regarding municipalities use of photo-radar in South Carolina. Though the opinion notes that no State statute prohibited photo-radar enforcement, the Attorney General nevertheless expressed concerns about the registered owner presumption, concluding that the Legislature was the appropriate authority to authorize use of the presumption.

3) Office of the Attorney General of the State of Montana, 45 Op. Atty Gen. Mont. 7 (1993) regarding a municipality enacting a photo-radar ordinance. The Attorney General's opinion noted "a presumption exists that legislative acts are constitutional" and "the constitutionality of a proposed legislative act is not an appropriate subject for an Attorney General's Opinion."
4) Office of the Attorney General of the State of Alabama, 239 Op. Atty Gen. Ala. 52 (1995) regarding the use of photo radar devices. The Attorney General indicated that, "while the use of such devices is legal, the use of such devices to mail speeding citation to motorists would not comply with substantive or procedural requirement of Alabama law."
5) Office of the Attorney General of the State of Georgia, No. 82000-7, 2000 Ga. AG LEXIS 13 (2000) concluding the "Home Rule Act" allowed municipalities to enact photo enforcement programs.
6) Office of the Attorney General of the State of Georgia, No. U2000-12, 2000 Ga. AG LEXIS 23 (2000) concluding counties may enact ordinances permitting photo enforcement and whether such devices may be used within the state highway system.
7) Andrew N. J. Tarr, Picture It: Red Light Cameras Abide by the Law of the Land, 80 N.C. L. REV., 1879 (2002).
8) Mark Lehman, Are Red Light Cameras Snapping Privacy Rights?, 33 U. TOL. L. REV., 815 (2002).
9) Steven Tafoya Naumchi, Review of Selected 1998 California Legislation, Transportation and Motor Vehicles: Stop Photographic Enforcement of Red Lights, 30 MCGEORGE L. REV., 833 (1999).
10) Thomas M. Stanek, Comment, Photo Radar in Arizona: Is it Constitutional?, 30 ARIZ ST. L.J., 1209 (1998).

AUTOMATED ENFORCEMENT RELATED STATUTES AND ORDINANCES

Model Statute:

National Committee of Uniform Traffic Laws and Ordinances, *Automated Traffic Law Enforcement Model Law* – www.ncutlo. org/autoenforce622.htm.

State Statutes:

1) California Vehicle Code– CAL. VEH. CODE §§ 210, 21455.5, 21455.6, 40518, 40520 (2003).
2) Colorado Revised Statutes – COLO. REV. STAT. § 42-4-110.5 (2002).
3) Delaware Code Annotated - DEL. CODE. ANN. TITL 21 § 4101(d) (2002).
4) Official Code of Georgia Annotated – GA. CODE. ANN. § 40-6-20 (2002).
5) Illinois Compiled Statutes Annotated - 625 ILL. COMP. STAT. ANN. 5/1-105.5, 5/11-306 (2002).
6) Annotated Code of Maryland – MD. CODE ANN. TRANSP. § 21-202.1 (2002).
7) Nevada Revised Statutes Annotated – NEV. REV. STAT. ANN. § 484.910 (2002).
8) New Jersey Annotated Statutes – *N.J. STAT. ANN. § 39:4-103.1 (2002).*
9) New York Consolidated Laws Service – N.Y. VEH. & TRAF. LAW § 1111-a (2002).
10) General Statutes of North Carolina – N.C. GEN. STAT. § 160A-300.1 (2002).
11) Oregon Revised Statutes – OR. REV. STAT. §§ *810.434* - 36, 438 - 439 (2001).
12) Pennsylvania Consolidated Statutes - 75 PA.C.S. §§ 102, 3116 (2002).
13) Utah Code Annotated –UTAH CODE ANN. § 41-6-52.5 (2002).
14) Code of Virginia – VA. CODE. ANN. §§ 46.2-819.1, 833.01 (2002).
15) Wisconsin Statutes - WIS. STAT. § 349.02 (2002).

Ordinances:

1) TOLEDO, OHIO, MUN. CODE § 313.12 (1999) and ORDINANCE NO. 451-00 (2000).
2) DAYTON, OHIO, REV. CODE OF GEN. ORDINANCES NO. 70.121 (2002).
3) DISTRICT OF COLUMBIA CODE ANN. §§ 50-2209.01, 03 (2002).
4) CHARLOTTE, N. C., ORDINANCE NO. 966 (1998)– see www.charmeck.org/Departments/ransportation/special+programs/city +ordinance.asp

REFERENCES

[1] Guidance for Using Red Light Cameras, Publication No. FHWA-SA-03-018, Federal Highway Administration, National Highway Traffic Safety Administration, Washington, DC, March 2003.

[2] Hugh W. McGee and Kimberly A. Eccles, "Impact of Red Light Camera Enforcement on Crash Experience", NCHRP Synthesis 310, National Cooperative Highway Research Program, Transportation Research Board, Washington, DC, 2003.

[3] F. M. Council, B. Persaud, K. Eccles, and C. Lyon, "Safety Evaluation of Red Light Cameras" FHWA Publication No., FHWA-HRT-05-048, November 2004.

[4] Federal Highway Administration, "Association of Selected Intersection Factors with Redlight-running Crashes," ITE Journal, Vol. 70, No. 7, July 2000, pp. 37-42.

[5] R. A. Retting, A.F. Williams, D.F. Preusser and H.B. Weinstein, "Classifying Urban Crashes for Countermeasure Development," Accident Analysis and Prevention, Vol. 27, No. 3, 1995, pp. 283-294.

[6] J. M. Golob, S. Cho, J.P. Curry and T.F. Golob, Impacts of the San Diego Photo Red-light Enforcement System on Traffic Safety, Transportation Research Board Pre-print CD-ROM, Washington, DC, January 2003.

[7] Making Intersections Safer: A Toolbox of Engineering Countermeasures to Reduce Red Light Running, Federal Highway Administration and Institute of Transportation Engineers, Washington, DC, 2003.

[8] As cited by Insurance Institute of Highway Safety at http://www. iihs.org/safety_facts/qanda/ rlc.htm.

[9] Determining Vehicle Signal Change and Clearance Intervals, An Informational Report of the Institute of Transportation Engineers, Prepared by ITE Technical Council Task Force 4TF1ITE, Washington, DC, August 1994.

[10] B.E. Porter and K. J. England, "Predicting Red-Light Running Behavior: A Traffic Safety Study in Three Urban Settings," Journal of Safety Research, Vol. 31, No. 1, pp 1-8, 2000.

[11] R.A. Retting and M.A. Greene, "Influence of Traffic Signal Timing on Red Light Running and Potential Vehicle Conflicts at Urban Intersections," Transportation Research Record 1595, Transportation Research Board, Washington, DC, 1997, pp. 1-7.

[12] F.R. Hulscher, "The Problem of Stopping Drivers after the Termination of the Green Signal at Traffic Lights," Traffic Engineering and Control, Vol. 25, No. 3, March 1984, pp. 110-116.
[13] Traffic Control Systems Handbook, Publication No. FHWA-SA-95-032, Federal Highway Administration, Washington, DC, 1995.
[14] James H. Kell and Iris J. Fullerton Eds., Manual of Traffic Signal Design, 2nd Ed., Institute of Transportation Engineers/Prentice Hall, Washington, DC, 1998.
[15] L. Evans and R.W. Rothery, "Influence of Vehicle Size and Performance on Intersection Saturation Flow," Proceedings of the Eighth International Symposium on Transportation and Traffic Theory, University of Toronto Press, Toronto, Ontario, 1983.
[16] Intersection Safety Issue Briefs, Federal Highway Administration and Institute of Transportation Engineers, Washington, DC, April 2004. http://www.ite.org/library/IntersectionSafety/BreifingSheets.pdf
[17] Manual on Uniform Traffic Control Devices for Streets and Highways, 2003 Edition, Federal Highway Administration, Washington, DC, 2003.
[18] H. Douglas Robertson, Ed., Manual of Transportation Engineering Studies, Institute of Transportation Engineers, Washington, DC, 2000.
[19] H. S. Stein, "Traffic Signal Change Intervals: Policies, Practices and Safety," Transportation Quarterly, Vol. 40, No. 3, 1986, pp.433-445.
[20] T.M. Tarawneh, V.A. Singh, and P.T. McCoy, "Investigation of Effectiveness of Media Advertising and Police Enforcement in Reducing Red Light Violations," Transportation Research Record 1693, Transportation Research Board, National Research Council, Washington, DC, 1999, pp. 37-45.
[21] B.B. Farraher, R. Weinholzer, and M. P. Kowski, The Effect of Advanced Warning Flashers on Red Light Running: A Study Using Motion Imaging Recording System Technology at Trunk Highway 169 and Pioneer Trail in Bloomington, Minnesota. In 69th ITE Annual Meeting, Institute of Transportation Engineers, Washington, DC, 1999.
[22] A Policy on Geometric Design of Highways and Streets, American Association of State Highway Transportation Officials, Washington, DC, 2001.
[23] The Traffic Safety Toolbox: A Primer on Traffic Safety, Institute of Transportation Engineers, Washington, DC, 1999.
[24] James L. Pline, Ed., Traffic Engineering Handbook, 5th Ed., Institute of Transportation Engineers, Washington, DC, 1999.

[25] Toolbox on Intersection Safety and Design, Institute of Transportation Engineers and Federal Highway Administration, Washington, DC, September 2004.
[26] National Committee of Uniform Traffic Laws and Ordinances, Automated Traffic Law Enforcement Model Law. www.ncutlo. org/autoenforce622.htm.
[27] John F. McFadden and Hugh W. McGee, Synthesis and Evaluation of Red Light Running Automated Enforcement Programs in the United States, Publication No. FHWA-A-IF-00-004, Federal Highway Administration, Washington, DC, 1999.

BIBLIOGRAPHY

Agent, K.R. and D. Wagner, *Evaluation of Red Light Running Campaign*, Report KTC-96-29, Kentucky Transportation Center, December 1996, pp. 1-39.

Automated Enforcement in Transportation, Institute for Transportation Engineers, Washington, DC, December 1999.

"Beep, Smash, Thud: Traffic Engineering Methods to Reduce Urban Crashes," *Status Report, Insurance Institute for Highway Safety*, Vol. 33, No. 4, May 2, 1998, pp. 1–7.

Blackburn, R.R. and D.T. Gilbert, "Photographic Enforcement of Traffic Laws," *National Cooperative Highway Research Program: Synthesis of Highway Practice 219,* Transportation Research Board, Washington, DC, 1995, pp. 1-68.

Bonneson, J. and Zimmerman, K. "Effect of Yellow Interval Timing On Red-Light Violation Frequency at Urban Intersections," Texas Transportation Institute, The Texas A & M University, *Transportation Research Record*, Washington, D.C., 2004.

Burris, M. and R. Apparaju, *Assessment of Automated Photo Enforcement Systems for Polk County Signalized Intersection Safety Improvement Project*, Center for Urban Transportation Research, University of South Florida, Tampa, Florida, September 25, 1998.

Butler, J.A., "Another View on Vehicle Change Intervals," *ITE Journal*, Vol. 53, No. 3, March 1983, pp. 44-48.

California State Auditor, *Red-light Camera Programs: Although They Have Contributed to a Reduction in Accidents, Operational Weaknesses Exist at the Local Level*, Bureau of State Audits, Sacramento, California, July 2002.

Chin, H. C., "Effect of Automatic Red-light Cameras on Red-Light Running," *Traffic Engineering and Control*, Vol. 30, No. 4, April 1989, pp. 175–179.

City of Charlotte, North Carolina, http://www.charmeck.org/Departments/ transportation/ special+programs/city+ ordinance.asp.

City of Toledo, Ohio. http://safety.fhwa.dot.gov/fourthlevel/srlr/ pdf/ Leg_Ohio.pdf

Curry, James, "Enforcement and Technology Innovations", *Proceedings of the 1997 International Highway-Rail Grade Crossing Safety Conference,* Conference, Seattle, Washington, July 1997.

Datta, T. K., K. Schattler, and S. Datta, "Red Light Violations and Crashes at Urban

Intersections," *Transportation Research Record 1734,* Paper No. 00-0480, January 2000, pp. 1-26.

Erickson, G., "Digital Traffic Cameras," *Traffic Technology International*, 1996, pp. 332–336.

Fleck, J. L., and B. B. Smith, "Can We Make Red Light Runners Stop? Red Light Photo Enforcement in San Francisco, California," *Transportation Research Record 1693,*

Transportation Research Board, National Research Council, Washington, DC, 1999, pp. 46-49.

Frisch, Al, "LA's Metro - Innovations In Grade Crossing Safety", *Transit Policing*, Volume 3, Number 2, Fall 1993.

Gilbert, Daniel T., Nina J. Sines, and Brandon E. Bell, "Photographic Traffic Law Enforcement," *National Cooperative Highway Research Program Legal Research Digest, Number 36,* Transportation Research Board, Washington, DC, December, 1996.

Hansen, G. A., "Human Factors in Transportation: Use of Automated Traffic Enforcement

Technology to Modify Driving Behavior," In *Transportation Research Board 78th Annual Meeting*. Transportation Research Board, National Research Council, Washington, DC, January 1999, pp. 1–15.

Hasson, P., "Red Lights Mean Stop," Midwest Resource Center, Federal Highway Administration, 2000, http://mrc.fhwa.dot.gov/articles/ redlight.htm.

Hill, S. and Lindly, J. "Speed Red Light Running Prediction And Analysis." Department of Civil and Environmental Engineering, The University of Alabama, *Transportation Research Record*, Washington, D.C., 2004.

Houston-Galveston Area Council, *Greater Houston Metropolitan Safety Planning Program: 2004 Status Report*, February 17, 2004, pp.1-37.

Hubaud, Lou and Linda Meadow. "Los Angeles Metro Blue Line Enforcement Program", Final *Report, Safety of Highway-Railroad Grade Crossings, Research Needs Workshop,* Report Number DOT-VNTSC-FRA-95-12.2, January 1996.

Insurance Institute for Public Safety, "Red Light Cameras Yield Big Reductions in Crashes and Injuries", *Status Report,* Volume 36, No. 4, April 28, 2001.

Kendall, S. "Is Automate Enforcement Constitutional?", *Insurance Institute for Highway Safety,* 2004.

Kraus, E., and Quiroga, C. "Red Light Running Trends In Texas," Texas Transportation Institute, The Texas A & M University, *Transportation Research Record*, Washington, D.C., 2004.

Lawson, S.D., "Automatic Surveillance and Red-Light running: Potential for Camera Use and Accident Reduction at High-Risk Light-Controlled Junctions," *Traffic Engineering and Control,* January 1992, pp. 10-12.

Lawson, S.D., H.T. Morris, R.W. Hardy, and A.C. Howard, *Red Light Running and Surveillance Cameras – Policy Issues Related to Accident Reduction and Enforcement.*

Meadow, Linda, "Los Angeles Metro Blue Line Grade Crossing Safety Improvement Program," *Proceedings of the 1993 National Conference on Highway-Rail Safety,* St. Louis, Missouri. July 11-14, 1993.

Meadow, Linda, "Los Angeles Metro Blue Line Grade Crossing Safety Improvement Program", *Paper Presented at 1993 Rapid Transit Conference, American Public Transit Association,* Washington, DC, June 1993.

Meadow, Linda, "Los Angeles Metro Blue Line Light Rail Safety Issues", *Transportation Research Record 1433,* Transportation Research Board, National Research Council, Washington, DC, pp. 123-133.

Meadow, Linda and James Curry, "New Technologies for Highway-Railroad Grade Crossing Safety", *Proceedings of the Seventh National Conference on Light Rail Transit, Volume 2.* Baltimore, Maryland, November, 1995.

Meadow, Linda and James Curry, "Light Rail Transit Safety Issues", *Compendium of Technical Papers, Institute of Transportation Engineers, 47th District 6 Annual Meeting,* Portland, Oregon, July 1994.

Mitretek Systems, *Automated Enforcement of Traffic Signals; A Literature Review*, Prepared For U.S. Department of Transportation, Federal Highway Administration, Contract Number DTFH61- 00-C-00001, Washington, DC, August 13, 2001.

Mohamedshah, Y. M., L. W. Chen, and F. M. Council, *Association of Selected Intersection Factors with Red Light Running Crashes,* Federal Highway Administration, U.S. Department of Transportation, Washington, DC, May 2000, pp. 1–21.

National Campaign to Stop Red Light Running, http://www.stopredlightrunning.com.

"Officials Nationwide Give a Green Light to Automated Traffic Enforcement," *Status Report,* Insurance Institute for Highway Safety, Vol. 35, No. 3, March 2000.

O'Laughlin, John, "ITS Technology and Law Enforcement", *ITS Quarterly,* Volume 5, Number 2. Intelligent Transportation Society of America, Washington, DC, Summer 1997, pp. 51-55.

PB Farradyne, *Photo Enforcement System Review*, Prepared for the City of San Diego Police Department, January, 2002, http://www.sandiego.gov/police.

Passetti, K. A., and T. H. Hicks, Use of Automated Enforcement for Red Light Violations, Department of Civil Engineering, Texas A&M University, College Station, Texas, pp. 1–60.

Photocop Home Page. Photocop, Cedar Park, Texas. http://www. photocop.com.

Polk, A., "Electronic Enforcement of Traffic Laws: The Devil is in the Details," *ITS Quarterly*, Summer 1998, pp. 12–27.

Redelmeier, D. "Traffic-Law Enforcement and Risk of Death from Motor-Vehicle Crashes: Case-Crossover Study," *The Lancet,* Vol. 361, June 28, 2003, pp. 2177-2182.

Red Light Cameras Q&A: General, Insurance Institute for Highway Safety, Arlington, Virginia, http://www.hwysafety.org/qanda/qarlc.htm.

Red Light Literature, Photocop, Cedar Park, Texas. http://www.photocop.com/red-light.htm. Accessed August 4, 1999.

Stop Red Light Running Literature, Federal Highway Administration, U.S. Department of Transportation, Washington, DC, http://www.fhwa.dot.gov/stoprlr/camr/camrtech.htm. Accessed November 1999.

Supriyasilp, T, D. Turner, J. Lindly, E. Mansfield, and J. Howell, *Speed Sensitivity of a Red Light Camera Enforcement System*, Department of Civil and Environmental Engineering, The University of Alabama, Presented at 84th Annual Meeting of the Transportation Research Board, Washington, DC, January 2004.

Swali, L. N., and P. L. Belcher, "The Use of Cameras to Reduce Casualties in London," In *International Conference on Road Traffic Monitoring and*

Control, Institution of Electrical Engineers, London, England, 1992, pp. 28–31.

Turner, S. and A. E. Polk. "Overview of Automated Enforcement in Transportation," *ITE Journal*, June 1998, pp. 20–28.

U.S. Department of Transportation, *Accidents That Shouldn't Happen*, A Report of the Grade Crossing Safety Task Force to Secretary Federico Pena, March 1, 1996, pp. 4.

Wissinger, L. M., J. E. Hummer, and J. S. Milazzo II, "Using Focus Groups to Investigate the Issues Surrounding Red Light Running," *Transportation Research Board 79th Annual Meeting (Preprint CD-ROM),* Transportation Research Board, National Research Council, Washington, DC, January 2000, pp. 1–28.

World Report on Road Traffic Injury Prevention, World Health Organization/The World Bank, Washington, DC, April 2004, pp. 132-133.

Vinzant, J. C., and B. J. Tatro, *Evaluation of the Effects of Photo Radar Speed and Red Light Camera Technologies on Motor Vehicle Crash Rates,* Department of Police, City of Mesa, Arizona, March 1, 1999, pp. 2–7.

Yamada, A., and P. Salzberg, *A Review of Automated Traffic Enforcement Systems: A Report to Washington State Legislature,* Washington Traffic Safety Commission, December 1999.

Yee, B. M., "San Francisco's Red Light Camera Enforcement Program," In *Compendium of Technical Papers: The 67th ITE Annual Meeting*. Institute of Transportation Engineers, Washington, DC, 1997.

Zador, P., H. Stien, S. Shapiro, and P. Tarnoff," Effect of Clearance Interval Timing on Traffic Flow and Crashes at Signalized Intersections," *ITE Journal,* November 1985, pp. 36-39.

End Note For Appendix A.

1 See Office of the Attorney General of the State of South Carolina, 1996 S.C. AG LEXIS 54 (1996).

INDEX

D

E

F

G

H

I

J

K

L

M

N

O

P

Q

R

S

T

U

V

W

Y